MASONRY
INSTANT ANSWERS

MASONRY
INSTANT ANSWERS

Rochelle C. Jaffe

McGraw-Hill

New York   Chicago   San Francisco   Lisbon   London
Madrid   Mexico City   Milan   New Delhi   San Juan
Seoul   Singapore   Sydney   Toronto

The **McGraw·Hill** *Companies*

Copyright © 2004 by The McGraw-Hill Companies, Inc. All rights reserved. Printed in the United States of America. Except as permitted under the United States Copyright Act of 1976, no part of this publication may be reproduced or distributed in any form or by any means, or stored in a data base or retrieval system, without the prior written permission of the publisher.

1 2 3 4 5 6 7 8 9 0   DOC/DOC   0 9 8 7 6 5 4 3

ISBN 0-07-139515-6

The sponsoring editor for this book was Larry S. Hager and the production supervisor was Sherri Souffrance. It was set in Stone Sans by Lone Wolf Enterprises, Ltd.

Printed and bound by RR Donnelley.

McGraw-Hill books are available at special quantity discounts to use as premiums and sales promotions, or for use in corporate training programs. For more information, please write to the Director of Special Sales, McGraw-Hill Professional, Two Penn Plaza, New York, NY 10121-2298. Or contact your local bookstore.

 This book is printed on recycled, acid-free paper containing a minimum of 50% recycled, de-inked fiber.

To Bob, who makes it all possible.

ABOUT THE AUTHOR: ROCHELLE C. JAFFE

Ms. Jaffe is a licensed architect and a licensed structural engineer. She is also certified as a construction specifier (by the Construction Specification Institute) and as a special inspector of structural masonry (by the International Conference of Building Officials). As a Senior Project Consultant with NTH Consultants, Ltd., in Farmington Hills, Michigan, Ms. Jaffe specializes in investigation, evaluation, and rehabilitation of existing, deteriorated, and damaged masonry structures, and has over 20 years' experience in this field.

Ms. Jaffe is a member of the American Concrete Institute (ACI) and the American Society of Civil Engineers (ASCE). As an active member of The Masonry Society (TMS), she is a former member of the Board of Directors and former editor of the Journal, serves on the Design Practices Committee, and is current chairperson of the Architectural Practices Committee. Her active participation in the ACI/ASCE/TMS Masonry Standards Joint Committee, the group that authors the Building Code Requirements for Masonry Structures (ACI 530/ASCE 5/TMS 402) and the Specification for Masonry Structures (ACI 530.1/ASCE 6/TMS 602), includes serving as a former chair of the Stone Cladding Subcommittee.

Ms. Jaffe has authored or co-authored magazine articles and texts, as well as numerous technical reports related to investigations of masonry structures with material and architectural or structural deficiencies. She has spoken at a variety of seminars and other technical functions.

CONTENTS

INTRODUCTION

Masonry Instant Answers addresses questions that arise at the job site during the course of masonry construction. Materials and installation issues are explained in an easily grasped format that is filled with tables and figures. Both industry recommendations and code requirements for masonry construction and inspection of the construction are presented in this text, which can be carried to the project site.

Design issues, such as selection of materials or structural capacity of the masonry, are not discussed in this book since those topics are outside the scope. More importantly, decisions related to masonry design must have been made prior to the start of masonry construction.

The Glossary defines masonry terms that are used in this book. Acronyms for masonry-related organizations are not only explained, but contact information is also provided for those organizations where individuals are prepared to answer specific questions.

Rochelle C. Jaffe, SE, Ar, CCS, CSISM

ACKNOWLEDGMENTS

Many sources were relied upon to provide information and artwork for this book. However, the following organizations and individual provided significant contributions, which are gratefully acknowledged. Contact information for the organizations is provided in Acronyms at the back of the book.

ASTM International
Christine Beall
International Building Code
Mason Contractors Association of America
Masonry Standards Joint Committee
National Concrete Masonry Association
Portland Cement Association

LIST OF TABLES

LIST OF FIGURES

chapter

1

MASONRY UNITS

The types of masonry units used in construction offer large variations in appearance and physical properties, including strength, fire resistance, sound transmission, thermal resistance, water penetration resistance, and light transmission. The variation in available sizes and shapes of man-made and natural masonry units is astounding and outside the scope of this book.

This chapter discusses the masonry units that are most commonly used in masonry construction. The physical, chemical, dimensional, and visual properties of each masonry unit are governed by a standard produced by American Society for Testing and Materials (ASTM) International. Where appropriate, the property requirements of the relevant ASTM standard are provided herein. The test methods used to evaluate compliance of the masonry units with their relevant standard are also discussed in this chapter.

CLAY UNITS

Clay is the general term applied to the materials used to manufacture structural clay products. The three principal forms in which it occurs have different physical characteristics but similar chemical compositions:

- Surface clays that are sedimentary and found near the surface of the earth
- Shales that have been subjected to high pressures that hardened them

1

fastfacts

Solid units have cores, cells, or frogs that constitute no more than 25 percent of their gross cross-sectional area. Hollow units have a net cross-sectional area that is less than 75 percent of their gross cross-sectional area.

- Fire clays that occur at greater depths than either surface clays or shales and are usually mined

Clay masonry units are produced by mixing the finely ground clay with water, molding or forming it into the desired shape, drying it, and burning it. After the units are formed, they are dried in kilns at temperatures from 110 degrees F to 300 degrees F for 24 to 48 hours. After drying, they are burned in kilns that gradually raise the temperature up to a maximum of 1600 degrees F to 2400 degrees F. The burning process can take two to five days. The materials used to fabricate the clay unit, the fineness with which the materials are ground, the length of burning, and the temperature and rate of burning all affect the clay unit's properties of color, size variation, absorption, compressive strength, and modulus of rupture.

Clay Bricks

There are four ASTM standards that govern clay masonry bricks, as listed in Table 1.1. Building bricks are used in both structural and nonstructural applications where appearance is not important. These units are used for the interior, hidden wythe of multi-wythe brick walls or the exterior, exposed wythe of walls that are rarely seen or serve utilitarian purposes. Face bricks are used for structural or facing components of the wall or both. Both building bricks and face bricks are solid units, meaning that they have cores, cells, or frogs that con-

TABLE 1.1   ASTM Standards for Clay Masonry Bricks

ASTM Std. No.	Name ("Standard Specification for …")
C 62	Building Brick (Solid Masonry Units Made From Clay or Shale)
C 216	Facing Brick (Solid Masonry Units Made From Clay or Shale)
C 652	Hollow Brick (Hollow Masonry Units Made From Clay or Shale)
C 1088	Thin Veneer Brick Units Made From Clay or Shale

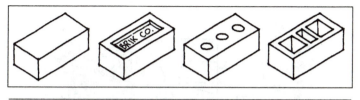

100% solid	frogged	cored < 25%	cored >25%
Solid Brick ASTM C 62 and ASTM C 216			Hollow Brick ASTM C 652

FIGURE 1.1   Solid and Hollow Clay Brick Units
(Beall and Jaffe, *Concrete and Masonry Databook,* McGraw-Hill, 2003).

stitute no more than 25 percent of their gross cross-sectional area. Hollow bricks may be building bricks or face bricks that have a net cross-sectional area that is less than 75 percent of their gross cross-sectional area. Hollow and solid clay bricks are illustrated in Figure 1.1. Thin veneer bricks, which are no more than 1¾ inches thick, are used to provide a non-structural surface finish to a wall.

All four of the clay brick standards provide for two or more grades of brick units. Brick grades, or weathering grades, differentiate the units according to their ability to resist damage caused by cyclic freezing. Definitions of the various brick grades are provided in Table 1.2. Weathering regions within the continental United States are shown in Figure 1.2. Table 1.3 lists recommendations for weathering

TABLE 1.2   Brick Grade Definitions

Grade	Definition
SW	severe weathering: high and uniform resistance to damage caused by cyclic freezing and brick may be frozen when saturated with water
MW	moderate weathering: moderate resistance to damage caused by cyclic freezing or brick may be damp but not saturated with water when freezing occurs
NW	negligible weathering: little resistance to damage caused by cyclic freezing in applications protected from water absorption and freezing
Exterior	exposed to weather
Interior	not exposed to weather

(Extracted, with permission, from ASTM C 62 *Standard Specification for Building Brick,* ASTM C 216 Standard Specification for Facing Brick, ASTM C 652 Standard Specification for Hollow Brick, and ASTM C 1088 *Standard Specification for Thin Veneer Brick Units Made From Clay or Shale,* copyright ASTM International, 100 Barr Harbor Drive, West Conshohocken, PA 19428. Copies of the complete standards may be obtained from ASTM, phone: 610-832-9585, email: service@astm.org, website: www.astm.org).

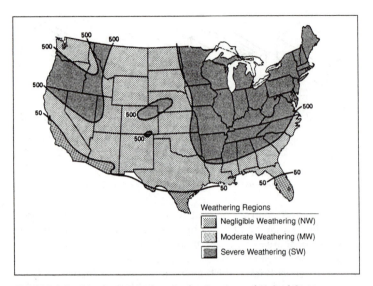

FIGURE 1.2 Weathering Regions in the Continental United States
(Extracted, with permission, from ASTM C 216 *Standard Specification for Facing Brick,*
copyright ASTM International, 100 Barr Harbor Drive, West Conshohocken, PA 19428.
A copy of the complete standard may be obtained from ASTM, phone: 610-832-9585,
email: service@astm.org, website: www.astm.org).

grades of bricks based on the weathering index of the region and
the intended use.

The physical property requirements for each type of brick unit,
including compressive strength, absorption, and efflorescence rating, vary according to the grade of the brick in each standard. Physical property requirements of the ASTM standards for clay bricks are
listed in Table 1.4.

TABLE 1.3 Clay Brick Grade Recommendations

Exposure In Exterior Walls	Weathering Index	
	Less than 50	50 and Greater
In vertical surfaces:		
In contact with earth	MW	SW
Not in contact with earth	MW	SW
In other than vertical surfaces:		
In contact with earth	SW	SW
Not in contact with earth	MW	SW

(Extracted, with permission, from ASTM C 216 *Standard Specification for Facing Brick,*
copyright ASTM International, 100 Barr Harbor Drive, West Conshohocken, PA 19428.
A copy of the complete standard may be obtained from ASTM, phone: 610-832-9585,
email: service@astm.org, website: www.astm.org).

TABLE 1.4 Physical Requirements of Clay Bricks

Clay Unit	Grade	Min. Compressive Strength, gross area (psi)		Max. Water Absorption by 5-hr Boiling (%)		Maximum Saturation Coefficient[a]	
		Average of 5 Tests	Individual Unit	Average of 5 Tests	Individual Unit	Average of 5 Tests	Individual Unit
Building brick[b] (ASTM C62)	SW	3000	2500	17.0	20.0	0.78	0.80
	MW	2500	2200	22.0	25.0	0.88	0.90
	NW	1500	1250	no limit	no limit	no limit	no limit
Face brick[b] (ASTM C216)	SW	3000	2500	17.0	20.0	0.78	0.80
	MW	2500	2200	22.0	25.0	0.88	0.90
Hollow brick[b] (ASTM C652)	SW	3000	2500	17.0	20.0	0.78	0.80
	MW	2500	2200	22.0	25.0	0.88	0.90
Thin brick[b] (ASTM C1088)	Exterior	—	—	17.0	20.0	0.78	0.80
	Interior	—	—	22.0	25.0	0.88	0.90

[a] The saturation coefficient is the ratio of absorption by 24-hr submersion in cold water to that after 5 hrs submersion in boiling water. The saturation coefficient requirement is waived when the 24-hr cold water absorption of each unit in a random 5-unit sample does not exceed 8%.
[b] When tested for efflorescence, the rating shall be "not effloresced".

(Extracted, with permission, from ASTM C 62 *Standard Specification for Building Brick*, ASTM C 216 *Standard Specification for Facing Brick*, ASTM C 652 *Standard Specification for Hollow Brick*, and ASTM C 1088 *Standard Specification for Thin Veneer Brick Units Made From Clay or Shale*, copyright ASTM International, 100 Barr Harbor Drive, West Conshohocken, PA 19428. Copies of the complete standards may be obtained from ASTM, phone: 610-832-9585, email: service@astm.org, website: www.astm.org.

For clay units in which the physical appearance is important, each ASTM standard also defines several different types of clay unit. Building bricks are not classified by type because appearance is not a requirement. Face bricks are classified by types FBS, FBX, and FBA; hollow bricks are classified by types HBS, HBX, HBA, and HBB; thin bricks are classified by types TBS, TBX, and TBA. Definitions of brick types are provided in Table 1.5. Permitted chippage, variation in size, and distortion of surfaces and edges vary with the brick type. The "BX" types have more stringent tolerances than the "BS" and "BB" types, while the "BA" types do not have limitations. When the project specifications do not list the required type of brick, the standard stipulates that the requirements for type FBS, HBS, or TBS (as applicable) govern.

All four standards for clay brick units limit the quantity of bricks that do not meet the standard's requirements for chippage and tolerances, including broken brick, to no more than 5 percent when delivered. The standards also list permitted variations in size and distortion as well as amount of chippage. These limitations apply to the as-delivered units. The standards for face brick (C 216), hollow brick (C 652), and thin brick (C 1088) have additional requirements for appearance of the faces that will be exposed after construction. These faces may not contain cracks or other imperfections that detract from the appearance when viewed from a distance of 15 feet for types FBX, HBX, and TBX, and from a distance of 20 feet for

TABLE 1.5 Clay Brick Types

Type[1]	Definition
FBS, HBS, TBS	for general use in masonry
FBX, HBX, TBX	for general use in masonry where a higher degree of precision and lower permissible variation in size than permitted for Type FBS is required
FBA, HBA, TBA	for general use in masonry; selected to produce characteristic architectural effects resulting from non-uniformity in size and texture of the individual units
HBB	for general use in masonry where a particular color, texture, finish, uniformity, or limits on cracks, warpage, or other imperfections detracting from the appearance are not a consideration

[1] When type is not specified for a project, Type FBS, HBS, or TBS shall govern.

(Extracted, with permission, from ASTM C 62 *Standard Specification for Building Brick*, ASTM C 216 *Standard Specification for Facing Brick*, ASTM C 652 *Standard Specification for Hollow Brick*, and ASTM C 1088 *Standard Specification for Thin Veneer Brick Units Made From Clay or Shale*, copyright ASTM International, 100 Barr Harbor Drive, West Conshohocken, PA 19428. Copies of the complete standards may be obtained from ASTM, phone: 610-832-9585, email: service@astm.org, website: www.astm.org).

types FBS, FBA, HBS, HBA, TBS, and TBA. There are no visual requirements for type HBB brick.

Glazed Bricks

The two ASTM standards that are applicable to glazed brick are:

- ASTM C 126 "Standard Specification for Ceramic Glazed Structural Clay Facing Tile, Facing Brick, and Solid Masonry Units"
- ASTM C 1405 "Standard Specification for Glazed Brick (Single Fired, Solid Brick Units)"

ASTM C 126 addresses the glaze coating on both hollow and solid units. The glaze may be applied prior to brick firing (single firing process) or may be applied after brick firing (double firing process). The standard does not include minimum criteria for durability of units exposed to exterior environments. ASTM C 1405 specifies strength and durability properties of both the coating and the clay brick unit, but only applies to solid units that are glazed in a single firing process. Physical requirements for ASTM C 1405 glazed units are based on the unit class (exterior or interior), while those for ASTM C 126 are based on the direction of the coring in the units. In ASTM C 1405, the class refers to the installation application of exterior or interior. When the class is not specified, the requirements for Class Exterior govern. Physical requirements for glazed clay units are given in Table 1.6.

Glazed clay units are specified by type, grade, and class. In both relevant ASTM standards, the type refers to the number of finished (glazed) faces. Type I units are for general use where only one finished face will be exposed. Type II units are for use where two opposite finished faces will be exposed. When the type is not specified, the requirements for Type I govern.

In ASTM C 126, Grade S (select) applies to units that will be placed with relatively narrow mortar joints, and Grade SS (select sized or ground edge) applies to units in which variation of face dimensions must be very small. In ASTM C 1405, however, Grade S (select) applies to units that will be placed with normal mortar joint widths (⅜ to ½ inch) and Grade SS (select sized or ground edge) applies to units that will be placed with narrow mortar joints. If the grade is not specified, the requirements for Grade S govern in both standards.

The standards list permitted variations in size and distortion as well as chips. These tolerances apply to the as-delivered units. Unless

TABLE 1.6 Physical Requirements of Glazed Units

ASTM Standard	Core Direction (C 126) or Class (C 1405)	Minimum Compressive Strength, gross area (psi)		Maximum Water Absorption by 24-hr Cold (%)		Maximum Saturation Coefficient[a]	
		Average of 5 Tests	Individual Unit	Average of 5 Tests	Individual Unit	Average of 5 Tests	Individual Unit
C 126	Vertical	3000	2500	—	—	—	—
	Horizontal	2000	1500	—	—	—	—
C 1405[c]	Exterior	6000	5000	—	7.0	0.78[b]	0.80[b]
	Interior	3000	2500	—	—	—	—

[a] The saturation coefficient is the ratio of absorption by 24-hr submersion in cold water to that after 5 hrs submersion in boiling water.
[b] The saturation coefficient requirement does not apply when the 24-hr cold water absorption of each unit does not exceed 6% and the average compressive strength of a random sample of 5 bricks equals or exceeds 8000 psi with no individual strength less than 7500 psi.
[c] When tested for efflorescence, the rating shall be "not effloresced".

(Extracted, with permission, from ASTM C 126 *Standard Specification for Ceramic Glazed Structural Clay Facing Tile, Facing Brick, and Solid Masonry Units,* and ASTM C 1405 *Standard Specification for Glazed Brick,* copyright ASTM International, 100 Barr Harbor Drive, West Conshohocken, PA 19428. Copies of the complete standards may be obtained from ASTM, phone: 610-832-9585, email: service@astm.org, website: www.astm.org).

otherwise agreed upon between the purchaser and the seller, the maximum amount of chipped, cracked, or broken glazed units in a delivery of ASTM C 126 glazed units is 3 percent. For ASTM C 1405 glazed bricks, broken units and those that do not meet the standard's requirements for chippage and size are limited to 5 percent of the delivery. The glazed faces that are exposed in the construction are further required to be free of chips, crazes, blisters, crawling, or other imperfections that detract from the appearance of the finished wall when viewed from a distance of 5 feet for ASTM C 126 glazed units and interior class ASTM C 1405 glazed units. This distance is increased to 15 feet for exterior class ASTM C 1405 glazed units.

Clay Pavers

There are three ASTM standards that cover various applications of clay brick pavers. The full titles of these standards are given in Table 1.7. ASTM C 410 paving bricks are for interior use in industrial applications ranging from food processing to airport terminals to chemical manufacturers. ASTM C 902 paving bricks are used in exterior applications for pedestrian traffic and low volumes of vehicular traffic, such as residential driveways and streets and commercial driveways (passenger drop-offs). ASTM C 1272 paving bricks are suitable for high volumes of heavy vehicles (trucks with three or more axles). Typical applications are streets, commercial driveways, and aircraft taxiways.

Each paver standard defines the paver categories differently. ASTM C 410 provides for four types of pavers: type T, type H, type M, and type L. These paver types are defined in Table 1.8. Physical and chemical requirements of ASTM C 410 pavers are given in Table 1.9.

ASTM C 902 defines pavers by their weather resistance (class SX, MX, or NX), traffic resistance (type I, II, or III), and dimensional tolerances (applications PS, PX, or PA). Classes, types, and applications for pedestrian and light traffic clay brick pavers are defined in Table 1.10. Dimensional tolerances are tighter for application PX than application PS, and no limits are given for application PA. Physical

TABLE 1.7   ASTM Standards for Clay Brick Pavers

ASTM Std. No.	Name ("Standard Specification for . . . ")
C 410	Industrial Floor Brick
C 902	Pedestrian and Light Traffic Paving Brick
C 1272	Heavy Vehicular Paving Brick

TABLE 1.8   ASTM C 410 Industrial Floor Brick Types

Type	Definition
T	for use where a high degree of resistance to thermal and mechanical shock is required but low absorption is not required
H	for use where resistance to chemicals and thermal shock are service factors but low absorption is not required
M	for use where low absorption is required; normally characterized by limited mechanical (impact) shock resistance but often highly resistant to abrasion
L	for use where minimal absorption and high degree of chemical resistance are required; normally characterized by very limited thermal and limited mechanical (impact) shock resistance but highly resistant to abrasion

(Extracted, with permission, from ASTM C 410 *Standard Specification for Industrial Floor Brick,* copyright ASTM International, 100 Barr Harbor Drive, West Conshohocken, PA 19428. A copy of the complete standard may be obtained from ASTM, phone: 610-832-9585, email: service@astm.org, website: www.astm.org).

requirements for pedestrian and light traffic paving bricks are given in Table 1.11 and abrasion resistance requirements are provided in Table 1.12.

Unless otherwise agreed upon between the purchaser and the seller, the maximum amount of units that do not meet the requirements of the standard for dimensional tolerance, distortion tolerance, and chippage in a delivery of ASTM C 902 brick pavers is 5 percent. The brick paving units must be free of cracks or other imperfections detracting from the appearance when viewed from a

TABLE 1.9   Physical and Chemical Requirements of ASTM C 410 Industrial Floor Bricks

Type	Minimum Modulus of Rupture (brick flatwise), gross area (psi)		Maximum Water Absorption by 5-hr Boiling (%)		Maximum Mass Loss by Chemical Resistance Test (%)
	Average of 5 Tests	Individual Unit	Average of 5 Tests	Individual Unit	
T	1000	750	10.0	12.0	no limit
H	1000	750	6.0	7.0	20.0
M	2000	1500	2.0	2.5	no limit
L	2000	1500	1.0	1.5	8.0

(Extracted, with permission, from ASTM C 410 *Standard Specification for Industrial Floor Brick,* copyright ASTM International, 100 Barr Harbor Drive, West Conshohocken, PA 19428. A copy of the complete standard may be obtained from ASTM, phone: 610-832-9585, email: service@astm.org, website: www.astm.org).

TABLE 1.10 ASTM C 902 Pedestrian and Light Traffic Paving Brick
Classes, Types, and Applications

Designation	Definition
Class SX	for use where the brick may be frozen while saturated with water
Class MX	for exterior use where resistance to freezing is not a factor
Class NX	not for exterior use, but may be acceptable for interior use where protected from freezing when wet
Type I	subject to extensive abrasion, such as driveways in publicly occupied spaces
Type II	subject to intermediate abrasion, such as residential driveways
Type III	subject to low abrasion, such as patios in single-family homes
Application PS	for general use and installed with mortar between units, or installed without mortar joints when laid in running or other bond not requiring extremely close dimensional tolerances
Application PX	for installation without mortar joints between units where exceptionally close dimensional tolerances are required as result of special bond pattern or unusual construction requirements
Application PA	for use where characteristic architectural effects resulting from non-uniformity in size, color, and texture are desired

(Extracted, with permission, from ASTM C 902 *Standard Specification for Pedestrian and Light Traffic Paving Brick,* copyright ASTM International, 100 Barr Harbor Drive, West Conshohocken, PA 19428. A copy of the complete standard may be obtained from ASTM, phone: 610-832-9585, email: service@astm.org, website: www.astm.org).

distance of 15 feet for application PX and a distance of 20 feet for application PS.

ASTM C 1272 defines heavy vehicular paving brick by type (based on intended installation) and application (dimensional tolerances, distortion, and chips). Definitions of designations used in ASTM C 1272 are provided in Table 1.13. Similar to ASTM C 902 pedestrian paving brick, permitted dimensional tolerances, distortion, and chippage are tighter for application PX than application PS, and no limits are given for application PA in ASTM C 1272. Type R pavers are intended to be used in a rigid paving applications, whereas Type F pavers are for use in flexible paving. Because the physical demands on pavers set in flexible paving are higher than

TABLE 1.11   Physical Requirements for ASTM C 902 Pedestrian and Light Traffic Paving Brick[a]

Unit Class	Minimum Compressive Strength (flatwise), gross area (psi)		Maximum Cold water Absorption (percent)		Maximum Saturation Coefficient[b]	
	Average of 5 Brick	Individual Brick	Average of 5 Brick	Individual Brick	Average of 5 Brick	Individual Brick
SX	8000	7000	8.0	11.0	0.78	0.80
MX	3000	2500	14.0	17.0	no limit	no limit
NX	3000	2500	no limit	no limit	no limit	no limit

[a] When tested for efflorescence, the rating shall be "not effloresced".
[b] The saturation coefficient is the ratio of absorption by 24-hr submersion in cold water to that after 5 hrs submersion in boiling water. The saturation coefficient requirement is waived if the average water absorption is less than 6% after 24-hr submersion in room-temperature water.

(Extracted, with permission, from ASTM C 902 *Standard Specification for Pedestrian and Light Traffic Paving Brick,* copyright ASTM International, 100 Barr Harbor Drive, West Conshohocken, PA 19428. A copy of the complete standard may be obtained from ASTM, phone: 610-832-9585, email: service@astm.org, website: www.astm.org).

those for pavers set in rigid paving, the physical requirements for Type F pavers are more stringent. Physical requirements for heavy vehicular paving bricks are given in Table 1.14 and abrasion resistance requirements are provided in Table 1.15.

Unless otherwise agreed upon between the purchaser and the seller, the maximum amount units of that do not meet the requirements of the standard for dimensional tolerance, distortion toler-

TABLE 1.12   Abrasion Resistance Requirements for ASTM C 902 Pedestrian and Light Traffic Paving Brick[a]

Abrasion Unit Type	Maximum Abrasion Index[b]	Max. Volum Loss (cm³/cm²)[c]
Type I	0.11	1.7
Type II	0.25	2.7
Type III	0.50	4.0

[a] Brick pavers shall meet the requirements of either the abrasion index or the volume abrasion loss. Values listed shall not be exceeded by any individual unit within the sample.
[b] Abrasion index is calculated from the cold absorption in percent and the compressive strength in psi as abrasion index = (100)(absorption)/compressive strength.
[c] Volume abrasion loss is determined from a modified version of ASTM C 418 Test Method for Abrasion Resistance of Concrete by Sandblasting.

(Extracted, with permission, from ASTM C 902 *Standard Specification for Pedestrian and Light Traffic Paving Brick,* copyright ASTM International, 100 Barr Harbor Drive, West Conshohocken, PA 19428. A copy of the complete standard may be obtained from ASTM, phone: 610-832-9585, email: service@astm.org, website: www.astm.org).

TABLE 1.13   ASTM C 1272 Heavy Vehicular Paving Brick Types and Applications

Designation	Definition
Type R	intended to be set in a mortar setting bed supported by an adequate concrete base; or an asphalt setting bed supported by an adequate asphalt or concrete base
Type F	intended to be set in a sand setting bed, with sand joints, and supported by an adequate base
Application PS	for general use
Application PX	for use where dimensional tolerances, warpage, and chippage are limited
Application PA	intended to produce characteristic architectural effects resulting from nonuniformity in size, color, and texture

ance, and chippage in a delivery of ASTM C 1272 brick pavers is 5 percent, including broken bricks. The brick paving units must be free or cracks or other imperfections, excluding chips, that detract from the appearance when viewed from a distance of 20 feet.

Clay Unit Testing

Only one ASTM standard is referenced for evaluation of compliance of clay masonry units with their pertinent standard: ASTM C 67 "Standard Test Methods for Sampling and Testing Brick and Structural Clay Tile." As stated in the title, the standard applies both to clay brick units and structural clay tile units. Many test procedures are described in this standard, not all of which are applicable to every type of clay unit. This chapter will summarize the most common testing that is relevant to clay brick units. However, the reader should refer to the ASTM standard for exact methods of testing. Additional test methods to evaluate the glaze on ASTM C 126 and ASTM C 1405 units are provided within those standards.

For lots that are up to 1,000,000 units in size, at least ten individual brick units are required to be tested for modulus of rupture, compressive strength, abrasion resistance, and absorption. For larger lots, five additional specimens must be selected for testing from each additional 500,000 bricks or fraction thereof.

Five dry, full-size units are tested for modulus of rupture (flexure). The bricks are tested flatwise, with a single point central loading.

TABLE 1.14 Physical Requirements for ASTM C 1272 Heavy Vehicular Paving Brick[a]

Unit Class	Minimum Thickness (in.)	Minimum Compressive Strength (flatwise), gross area (psi)		Maximum Cold water Absorption (percent)		Minimum Modulus of Rupture (psi)	
		Average of 5 Brick	Individual Brick	Average of 5 Brick	Individual Brick	Average of 5 Brick	Individual Brick
R	2¼	8000	7000	6.0	7.0	1200	1000
F	2⅜	10,000	8800	6.0	7.0	1500	1275

[a] When specified, units shall be tested for efflorescence (no requirement for efflorescence rating).

(Extracted, with permission, from ASTM C 1272 *Standard Specification for Heavy Vehicular Paving Brick*, copyright ASTM International, 100 Barr Harbor Drive, West Conshohocken, PA 19428. A copy of the complete standard may be obtained from ASTM, phone: 610-832-9585, email: service@astm.org, website: www.astm.org).

TABLE 1.15 Abrasion Resistance Requirements for ASTM C 1272 Heavy Vehicular Paving Brick[a]

Unit Type	Maximum Abrasion Index[b]	Max. Volume Abrasion Loss (cm^3/cm^2)[c]
Type R	0.11	1.7
Type F	0.11	1.7

[a] Brick pavers shall meet the requirements of either the abrasion index or the volume abrasion loss.
[b] Abrasion index is calculated from the cold absorption in percent and the compressive strength in psi as abrasion index = (100)(absorption)/compressive strength
[c] Volume abrasion loss is determined from a modified version of ASTM C 418 Test Method for Abrasion Resistance of Concrete by Sandblasting.

(Extracted, with permission, from ASTM C 1272 *Standard Specification for Heavy Vehicular Paving Brick,* copyright ASTM International, 100 Barr Harbor Drive, West Conshohocken, PA 19428. A copy of the complete standard may be obtained from ASTM, phone: 610-832-9585, email: service@astm.org, website: www.astm.org).

The reported modulus of rupture value is the average of the five test results.

Five dry, half-length units are tested for compressive strength. Units are tested flatwise, and bearing surfaces that are recessed or paneled must be capped before testing. The compressive strength of each test specimen is reported.

The scale or balance used to evaluate brick absorption is required to have a capacity of not less than 2000 g and to be sensitive to 0.5 g. Five half-length units are tested. The specimens are dried, cooled, and weighed prior to submersion and then re-weighed after submersion for the specified length of time. Both cold water and boiling water absorption are reported to the nearest 0.1 percent. The saturation coefficient (ratio of 24-hour cold water absorption to 5-hour boiling absorption) is reported to the nearest 0.01.

Freezing and thawing testing consists of cycles of 4 hour submersion in a 75 degree F water-filled thawing tank followed by 20 hours in ½ inch of water in a freezer. The five half-length bricks are subjected to 50 cycles unless withdrawn from the test due to disintegration. At the conclusion of the test, the units are reported as failing if any one of the following three conditions occur:

- Weight loss greater than 0.5 percent
- Specimen separates into two or more significant pieces
- Specimen develops a crack whose length exceeds the minimum dimension of the specimen

To evaluate the initial rate of absorption, five whole bricks are used. After weighing the dried unit, the unit is set on supports in a tray that has sufficient water to cover the bottom ⅛ inch of the brick unit. The brick and water level is maintained for one minute, after which the brick is removed, surface water is wiped off, and the brick is re-weighed. The reported result is the weight gain, corrected to a basis of 30 square inches of flatwise area.

Efflorescence testing is performed on ten whole bricks. Five specimens are placed on end in a tray that contains a 1-inch depth of distilled water for seven days. The other five specimens are stored in the same room but without contact with water. At the end of the week, the two sets of specimens are visually compared. The reported result is "effloresced" or "not effloresced."

Other test methods described in ASTM C 67 include: weight per unit area measurement of size, warpage, length change, void area in cored units; void area in deep frogged units; out-of-square, and shell and web thickness. These procedures generally require use of a steel rule, steel carpenter's square, calipers, and micrometer.

CONCRETE UNITS

Concrete units are made of the same ingredients as cast-in-place concrete: portland cement, graded aggregates, and water. To enhance properties or appearance, other ingredients may be added, such as air-entraining agent, water-repellent agent, coloring pigment, and pozzolanic materials.

The manufacturing process consists of machine-molding very dry, no-slump concrete into the desired shape. The molded shapes are then subjected to an accelerated curing procedure, either by heating the units in a steam kiln at atmospheric pressure at temperatures ranging from 120 to 180 degrees F for periods up to 18 hours or by subjecting the units to saturated steam at 325 to 375 degrees F in a large cylindrical pressure vessel for a period up to 12 hours after an initial preset period of 1 to 4 hours. The units are then dried.

Concrete Bricks

Concrete bricks are solid units, as shown in Figure 1.3. They may be made of concrete materials or of sand and lime. Two ASTM standards govern the manufacture of concrete bricks:

- ASTM C 55 "Standard Specification for Concrete Brick"
- ASTM C 73 "Standard Specification for Calcium Silicate Brick (Sand-Lime Brick)"

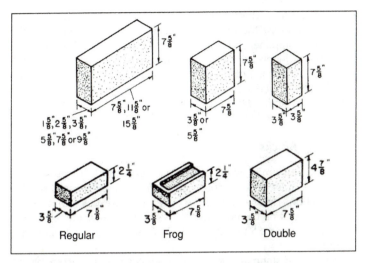

Regular   Frog   Double

FIGURE 1.3 Concrete Brick Units
(Panarese, Kosmatka, & Randall, *Concrete Masonry Handbook,* Portland Cement Association, 1991).

Both types of concrete bricks are designated by grade, but the grade definitions differ in the two standards. The concrete brick grade definitions are shown in Table 1.16.

The physical property requirements for concrete brick and sand-lime brick include minimum compressive strength and maximum

TABLE 1.16 Concrete Brick Grades

ASTM and Grade	Definition
C 55–N	for use as architectural veneer and facing units in exterior walls; for use where high strength and resistance to moisture penetration and severe frost action are desired
C 55–S	for general use where moderate strength and resistance to frost action and moisture penetration are required
C 73–SW	for use where exposed to temperature below freezing in the presence of moisture
C 73–MW	for use where exposed to temperature below freezing but unlikely to be saturated with water

(Extracted, with permission, from ASTM C 55 *Standard Specification for Concrete Brick* and ASTM C 73 *Standard Specification for Calcium Silicate Brick,* copyright ASTM International, 100 Barr Harbor Drive, West Conshohocken, PA 19428. Copies of the complete standards may be obtained from ASTM, phone: 610-832-9585, email: service@astm.org, website: www.astm.org).

water absorption. These requirements are shown in Table 1.17. The standards also list the permitted tolerances in dimensions and the allowable chippage and cracking. When units of either concrete brick or sand-lime brick are exposed in the wall construction, the exposed faces shall not show chips or cracks not otherwise permitted or other imperfections when viewed from a distance of not less than 20 feet under diffused lighting.

Concrete Blocks

Concrete block may be hollow or solid units. Solid concrete blocks have core holes that comprise no more than 25 percent of their cross-section. Hollow concrete blocks have core holes that reduce their cross-section by more than 25 percent. Hollow concrete blocks are usually about 55 percent solid. Concrete blocks are illustrated in Figure 1.4.

Two ASTM standards govern the manufacture of concrete blocks:

- ASTM C 90 "Standard Specification for Loadbearing Concrete Masonry Units"
- ASTM C 129 "Standard Specification for Nonloadbearing Concrete Masonry Units"

In both standards, the concrete blocks are classified according to their density, or unit weight. Lightweight units are those made of material having a density less than 105 pounds per cubic foot (pcf); medium weight units have a density of least 105 pcf up to 125 pcf; and normal weight units have a density of 125 pcf or more. The variation in unit density is accomplished by changing the type of aggregate used in the concrete mix. Table 1.18 demonstrates the differences in concrete unit density that can be achieved by changing the aggregate.

As shown in Table 1.19, the load-bearing units have a higher compressive strength requirement than the non-load-bearing units, and only the load-bearing units are required to meet the maximum water absorption requirements. The linear shrinkage of both types of concrete block units is limited to 0.065 percent at the time of delivery.

The ASTM standard for load-bearing concrete blocks also stipulates minimum thicknesses of face shells and webs in the concrete masonry units. Both standards list permitted size variations. Like concrete bricks, the exposed faces of both load-bearing and non-load-bearing concrete blocks in a wall shall not show chips or cracks

TABLE 1.17 Physical Requirements for ASTM C 55 Concrete Brick and ASTM C 73 Sand-Lime Brick

Weight Unit and Grade	Minimum Compressive Strength (psi)[a]		Maximum Water Absorption (pcf) of Average of 3 Units Based on Weight Classification[b]		
	Average of 3 Units	Individual Unit	Lightweight (<105 pcf)	Medium Weight (105 to 125 pcf)	Normal (>125 pcf)
Concrete Brick[c]					
Grade N	3500	3000	15.0	13.0	10.0
Grade S	2500	2000	18.0	15.0	13.0
Sand-Lime Brick[d]					
Grade SW	4500	3500	—	—	10.0
Grade MW	2500	2000	—	—	13.0

[a] tested flatwise, based on average gross area.
[b] based on over-dry weight of unit.
[c] At the time of delivery to the purchaser, the total linear drying shrinkage of concrete brick shall not exceed 0.065 percent.
[d] Sand-lime brick are not designated by weight classification.

(Extracted, with permission, from ASTM C 55 *Standard Specification for Concrete Brick* and ASTM C 73 *Standard Specification for Calcium Silicate Brick*, copyright ASTM International, 100 Barr Harbor Drive, West Conshohocken, PA 19428. Copies of the complete standards may be obtained from ASTM,

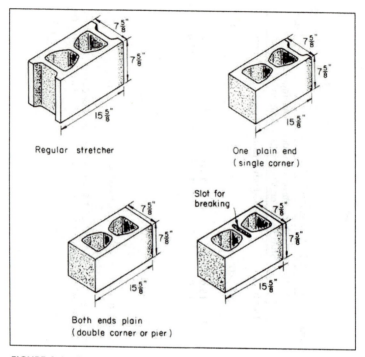

FIGURE 1.4 Concrete Blocks
(Panarese, Kosmatka, & Randall, *Concrete Masonry Handbook,* Portland Cement Association, 1991).

TABLE 1.18 Concrete Unit Weight Ranges

Concrete	Unit Weight, pcf
Sand and gravel concrete	130–145
Crushed stone and sand concrete	120–140
Air-cooled slag concrete	100–125
Coal cinder concrete	80–105
Expanded slag concrete	80–105
Pelletized-fly ash concrete	75–125
Scoria concrete	75–100
Expanded clay, shale, slate, and sintered fly ash concretes	75–90
Pumice concrete	60–85
Cellular concrete	25–44

(Panarese, Kosmatka, & Randall, *Concrete Masonry Handbook,* Portland Cement Association, 1991).

TABLE 1.19 Physical Requirements for Concrete Block Units

| Weight Unit | Minimum Compressive Strength (psi)[a] | | Maximum Water Absorption (pcf) of Average of 3 Units Based on Weight Classification[b] | | |
	Average of 3 Units	Individual Unit	Lightweight (<105 pcf)	Medium Weight (105 to 125 pcf)	Normal (>125 pcf)
Loadbearing CMU (ASTM C 90)	1900	1700	18.0	15.0	13.0
Nonloadbearing CMU (ASTM C 129)	600	500	—	—	—

[a] based on net cross-sectional area.
[b] based on over-dry weight of unit.

(Extracted, with permission, from ASTM C 90 Standard Specification for Loadbearing Concrete Masonry Units and ASTM C 129 Standard Specification for Nonloadbearing Concrete Masonry Units, copyright ASTM International, 100 Barr Harbor Drive, West Conshohocken, PA 19428. Copies of the complete standards may be obtained from ASTM, phone: 610-832-9585, email: service@astm.org, website: www.astm.org).

not otherwise permitted or other imperfections when viewed from a distance of not less than 20 feet under diffused lighting.

Specialty concrete block units must also meet the applicable ASTM standard for physical properties and tolerances. Examples of specialty block units, shown in Figure 1.5, are architectural blocks with special shaped exposed faces and structural blocks that accommodate placement of horizontal reinforcing bars.

Prefaced Concrete Units

Concrete and calcium silicate (sand-lime) masonry units may be covered at the point of manufacture with a smooth resinous facing. The facing may be resin, resin and inert filler, or cement and inert filler. The units are manufactured so that the surfaces that will be exposed to view in the final construction are the prefaced surfaces.

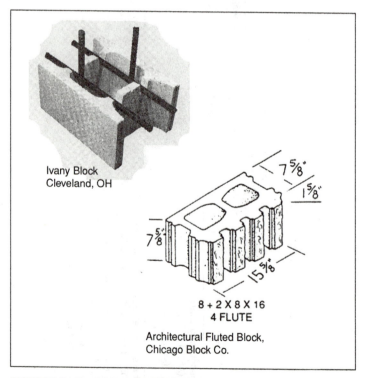

Ivany Block
Cleveland, OH

$7\frac{5}{8}"$

$1\frac{5}{8}"$

$7\frac{5}{8}"$

$15\frac{5}{8}"$

8 + 2 X 8 X 16
4 FLUTE

Architectural Fluted Block,
Chicago Block Co.

FIGURE 1.5   Special Concrete Block Shapes

ASTM C 744 "Standard Specification for Prefaced Concrete and Calcium Silicate Masonry Units" governs the manufacture of the facing only; properties of the concrete masonry unit body are regulated by specifying the relevant ASTM standard for the unit. The facing standard defines required properties and characteristics of the facing. It also lists the various chemicals to which the facing should be resistant for the time period specified, as shown in Table 1.20.

Manufacturing tolerances for size, distortion, cracking, and chippage are more stringent than those specified in the ASTM standards for the concrete units themselves. For example, the total variation from specified dimension of the facing must not exceed $\frac{1}{16}$ inch, while the total variation from the specified dimension of the underlying concrete unit can be $\frac{1}{8}$ inch. The appearance of the coating surface when placed in the wall is also more critical. That surface must be free of chips, crazes, cracks, blisters, crawling, holes, and other imperfections that detract from the appearance of the units when viewed from a distance of 5 feet perpendicular to the coated surface and using daylight without direct sunlight.

Concrete Pavers

Concrete paving units are governed by one of two ASTM standards:

- ASTM C 936 "Standard Specification for Solid Concrete Interlocking Paving Units"

TABLE 1.20 ASTM C 744 Chemical Resistance Requirements

Chemical	Test Duration (hrs).
Acetic acid (5%)	24
Hydrochloric acid (10%)	3
Potassium hydroxide (10%)	3
Trisodium phosphate (5%)	24
Hydrogen peroxide (3%)	24
Household detergent (10%)	24
Vegetable oil	24
Blue-black ink	1
Ethyl alcohol (95%)	3

(Extracted, with permission, from ASTM C 744 *Standard Specification for Prefaced Concrete and Calcium Silicate Masonry Units*, copyright ASTM International, 100 Barr Harbor Drive, West Conshohocken, PA 19428. A copy of the complete standard may be obtained from ASTM, phone: 610-832-9585, email: service@astm.org, website: www.astm.org).

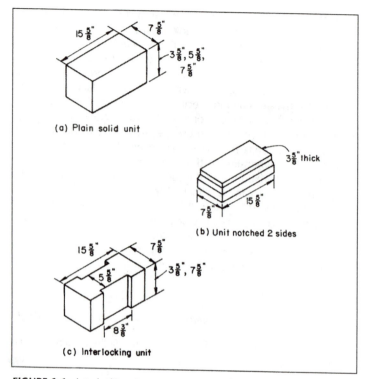

(a) Plain solid unit

(b) Unit notched 2 sides

(c) Interlocking unit

FIGURE 1.6 Interlocking Concrete Paver Units
(Panarese, Kosmatka, & Randall, *Concrete Masonry Handbook,* Portland Cement Association, 1991).

- ASTM C 1319 "Standard Specification for Concrete Grid Paving Units"

Interlocking pavers are relatively small units that are used to construct paved surfaces. They are required to be able to be lifted and placed with one hand. The exposed face cannot exceed 100.75 square inches, and the ratio of length to thickness is limited to 4. The minimum thickness is 2⅜ inches. Interlocking pavers are shown in Figure 1.6, and the physical requirements for these pavers are given in Table 1.21. For durability, the standard requires freeze-thaw testing or satisfactory field performance.

Grid pavers are used in vehicular trafficways, parking areas, soil stabilization, and revetments. The units are relatively large, with maximum face dimensions of 24 inches by 24 inches and minimum nom-

TABLE 1.21 Physical Requirements for ASTM C 936 Concrete Interlocking Paving Units

Compressive Strength (psi)		Absorption percent		Freeze-Thaw Resistance, Dry Mass Loss in 50 Cycles (%)	Abrasion Resistance[a]		Dimensional Tolerance (in.)	
Avg.	Indiv. Unit	Avg.	Indiv. Unit		Volume Loss (in^3/7.75 in^2)	Thickness Loss (in.)	Length or Width	Height
8000	7200	5.0	7.0	1.0	0.915	0.118	1/16	1/8

[a] When tested in accordance with ASTM C 418 "Test Method for Abrasion Resistance of Concrete by Sandblasting"

(Extracted, with permission, from ASTM C 936 *Standard Specification for Solid Concrete Grid Paving Units*, copyright ASTM International, 100 Barr Harbor Drive, West Conshohocken, PA 19428. A copy of the complete standard may be obtained from ASTM, phone: 610-832-9585, email: service@astm.org, website: www.astm.org).

fastfacts

ASTM C 1319 for Concrete Grid Pavers differs from most other ASTM standards for masonry units, in that it requires the units to maintain their physical properties for three years. Standards for other masonry materials specifically state that the physical property requirements do not apply to units that have been in service.

inal thickness of 3⅛ inches. Physical requirements for grid pavers are shown in Table 1.22. Durability of grid pavers is established by satisfactory field performance. Satisfactory field performance means that after three years in the same environment, temperature range, and traffic volume as the intended use, the pavers still satisfy the physical requirements of Table 1.22.

Concrete Unit Testing

One ASTM standard is referenced for evaluation of compliance of concrete masonry units with most properties in their pertinent standards: ASTM C 140 "Standard Test Methods for Sampling and Testing Concrete Masonry Units and Related Units." Several test procedures are described in this standard, and this chapter will summarize the most common testing that is relevant to concrete masonry units. However, the reader should refer to the ASTM standard for exact methods of testing. Additional test methods to evaluate the glaze on ASTM C 744 units are provided within that standard. ASTM C 426 "Standard Test Method for Linear Drying Shrinkage of Concrete Masonry Units" provides the procedure for evaluating the shrinkage potential of concrete units, which is a requirement in most concrete masonry material standards.

For lots that are up to 10,000 units in size, at least six individual concrete units are required to be sampled for compressive strength, absorption, unit weight (density), and moisture content in accordance with ASTM C 140. Twelve units should be sampled from lots that are between 10,000 and 100,000 units in size. For lots of more than 100,000 units, six additional specimens must be selected for testing from each additional 50,000 units or fraction thereof.

For each sample of six units, three must be tested for compressive strength. ASTM C 140 provides requirements for the testing apparatus, including the bearing plates that contact the test specimen. Test

TABLE 1.22 Physical Requirements for ASTM C 1319 Concrete Grid Paving Units

Net Area Compressive Strength (psi)		Maximum Water Absorption (lb/cu ft)	Minimum Net Area (%)	Web Width (in.)		Dimensional Tolerance, Height, Width, or Length (in.)
Average of 3 Units	Individual Unit			Minimum	Average	
5000	4500	10	50	1.00	1.25	1/8

(Extracted, with permission, from ASTM C 1319 *Standard Specification for Concrete Grid Paving Units*, copyright ASTM International, 100 Barr Harbor Drive, West Conshohocken, PA 19428. A copy of the complete standard may be obtained from ASTM, phone: 610-832-9585, email: service@astm.org, website: www.astm.org).

accuracy can be compromised by using bearing plates that do not cover the full area of the test specimen and have insufficient thickness (stiffness). Test specimens must be capped with either sulfur and granular materials or with gypsum cement in order to produce uniform and parallel testing surfaces. At the time of testing, the specimens must be free of visible moisture or dampness. The maximum compressive load is reported. For brick units, the compressive strength is reported on the gross area and for block units, the compressive strength is reported on net cross-sectional area.

For concrete roof pavers, three full-size units are required to be tested for flexural (bending) capacity. When the top surface is irregular, the test specimen must be capped as is done for compression testing. The flexural testing applies a single line load along the center of the test specimen.

Absorption testing may be performed on full-size units or on specimens that have been saw-cut from full-size units. For absorption testing, the test specimens are immersed in 60 to 80 degree F water for 24 hours. The specimens are weighed while immersed, then removed from the water and allowed to drain for 1 minute. The saturated weight is then recorded. The specimens are then dried in a ventilated oven at 212 to 239 degrees F for not less than 24 hours and until two successive weighings at intervals of 2 hours show an increment loss of no more than 0.2 percent of the previously determined weight. This value is then recorded as the oven-dried weight.

ASTM C 140 also discusses the method of measuring unit dimensions, and provides calculation methods and report procedures. The appendix to that standard contains worksheets for that use.

STONE MASONRY UNITS

Dimensioned stone masonry units are manufactured by quarrying the stone out of the earth, cutting the pieces to size, and finishing the exposed surfaces. Rubble stone units are also quarried, but are minimally dressed for size. Stones are natural products rather than man-made. As such, they should be expected to exhibit a broader variation in appearance and properties.

Stone Types

Dimensioned stone masonry units that are used in construction are of five main types: marble, limestone, granite, quartz-based stone,

fastfacts

Stones are natural products rather than man-made. As such, they should be expected to exhibit a broader variation in appearance and properties.

and slate. Each type is governed by a different ASTM standard, as listed in Table 1.23.

ASTM C 503 defines the properties for marble that are used on building exteriors and for structural purposes. ASTM defines marble as a crystalline rock that is capable of taking a polish and is composed predominantly of one or more of the following minerals: calcite, dolomite, or serpentine. Marble is classified into four categories, based on density:

- I Calcite
- II Dolomite
- III Serpentine
- IV Travertine

ASTM C 503 requires that when the marble will be used in an exterior application, the marble must be sound, durable, and free of spalls, cracks, open seams, pits, or other defects that are likely to impact its structural integrity in its intended use.

ASTM C 568 defines the properties for limestone that is used for general building and structural purposes. ASTM defines limestone as a sedimentary rock composed principally of calcium carbonate (calcite), the double carbonate of calcium and magnesium (dolomite),

TABLE 1.23 ASTM Standards for Dimension Stone

ASTM Std. No.	Title ("Standard Specification for . . ")
C 503	Marble Dimension Stone (Exterior)
C 568	Limestone Dimension Stone
C 615	Granite Dimension Stone
C 616	Quartz-Based Dimension Stone
C 629	Slate Dimension Stone

or a mixture of the two. Limestone is classified into three categories, based on density:

- I Low Density (110 through 135 pounds per cubic foot [pcf])
- II Medium Density (greater than 135 but no more than 160 pcf)
- III High Density (greater than 160 pcf)

For granite, ASTM C 615 defines the properties of stone used for general building and structural purposes. Specific permitted uses of granite dimension stone include: exterior and interior cladding; curbstone, paving, and landscape features; structural components; grade separations and retaining walls; and monuments. ASTM defines granite as a visibly granular, igneous rock consisting mostly of quartz and feldspars, accompanied by one or more dark minerals.

ASTM C 616 defines the properties of quartz-based stone used for general building and structural purposes. Quartz-based stone consists of sandstone and variations of sandstone, and is classified according to the free silica content of the stone:

- I Sandstone
- II Quartzitic Sandstone
- III Quartzite

ASTM describes sandstone as a sedimentary rock composed mostly of mineral and rock fragments within the sand size range of 2 to 0.6 mm, having a minimum of 60 percent free silica cemented or bonded to a greater or lesser degree by various materials including silica, iron oxides, carbonates, or clay, and that fractures around (not through) the constituent grains. Quartzitic sandstone is sandstone that has at least 90 percent free silica and that fractures around or through the constituent grains. Quartzite is highly indurated, typically metamorphosed sandstone containing at least 95 percent free silica and that fractures through the constituent grains.

ASTM C 629 defines the properties of slate used for general building and structural purposes. It specifically excludes slate that is used for roofing and industrial purposes. ASTM describes slate as a microcrystalline metamorphic rock most commonly derived from shale and composed mostly of micas, chlorite, and quartz. The sub-parallel orientation of the minerals allows the stone to be split into thin but tough sheets. Slate is classified by use:

- I Exterior
- II Interior

In each of the respective standards for limestone, granite, quartz-based stone, and slate, ASTM requires the stone to be sound, durable, and free of spalls, cracks, open seams, pits, or other defects that are likely to impact its structural integrity in its intended use.

Stone Testing

A number of ASTM standards define the sampling and test methods for evaluating stone properties. The relevant standard numbers and titles are listed in Table 1.24.

For bulk gravity (density) testing in accordance with ASTM C 97, which applies to marble, limestone, granite, and quartz-based stone, at least three specimens are required from each sample. The specimens may be cubes, prisms, cylinders, or any regular form with minimum dimension 2 inches, maximum dimension 3 inches, and ratio of volume to surface area not less than 0.3 and not more than 0.5 when measuring in inches.

In absorption testing of marble, limestone, granite, and quartz-based stone, the same specimens used for specific bulk gravity (density) may be used for evaluation of absorption according to ASTM C 97. For absorption testing of slate, ASTM C 121 requires test specimens to be square or rectangular slabs from $\frac{3}{16}$ to $\frac{5}{16}$ inch thick and not less than 4 inches on any side. At least six slate specimens are required.

Marble, limestone, granite, and quartz-based stone test specimens for compressive testing, according to ASTM C 170, may be

TABLE 1.24 ASTM Standards for Testing Dimension Stone

ASTM Std. No.	Title ("Standard Test Methods for…")
C 97	Absorption and Specific Bulk Gravity of Dimension Stone
C 99	Modulus of Rupture of Dimension Stone
C 120	Flexure Testing of Slate (Modulus of Rupture, Modulus of Elasticity)
C 121	Water Absorption of Slate
C 170	Compressive Strength of Dimension Stone
C 217	Weather Resistance of Slate
C 241	Abrasion Resistance of Stone Subjected to Foot Traffic
C 880	Flexural Strength of Dimension Stone
C 1353	Using the Tabor Abraser for Abrasion Resistance of Dimension Stone Subjected to Foot Traffic

TABLE 1.25 Physical Requirements for Dimension Stone

Stone, Standard, and Classification	Maximum Absorption by Weight[a] (%)	Minimum Density[b] (%)	Minimum Compressive Strength[c] (psi)	Minimum Modulus of Rupture[d] (psi)	Minimum Abrasion Resistance (hardness)[e]	Minimum Flexural Strength[f] (psi)
Marble						
I Calcite	0.20	162	7500	1000	10	1000
II Dolomite	0.20	175	7500	1000	10	1000
II Serpentine	0.20	168	7500	1000	10	1000
IV Travertine	0.20	144	7500	1000	10	1000
Limestone						
I Low Density	12.0	110	1800	400	10	—
II Med. Density	7.5	135	4000	500	10	—
III High Density	3.0	160	8000	1000	10	—
Granite	0.40	160	19000	1500	—	—
Slate[g]						
I Exterior	0.25	—	—	9000–7200[h]	8	—
II Interior	0.45	—	—	7200–5500[h]	8	—

(continued)

TABLE 1.25 Physical Requirements for Dimension Stone (*continued*)

Stone, Standard, and Classification	Maximum Absorption by Weight[a] (%)	Minimum Density[b] (%)	Minimum Compressive Strength[c] (psi)	Minimum Modulus of Rupture[d] (psi)	Minimum Abrasion Resistance (hardness)[e]	Minimum Flexural Strength[f] (psi)
Quartz-Based						
I Sandstone	8.0	125	4000	350	2[i]	—
II Quartzitic Sandstone	3.0	150	10,000	1000	8	—
III Serpentine	1.0	160	20,000	2000	8	—

[a] per ASTM C 121 for slate; ASTM C 97 for marble, limestone, granite, and quartz-based stone; [b] per ASTM C 97;
[c] per ASTM C 170; [d] per ASTM C 120 for slate; ASTM C 99 for marble, limestone, granite, and quartz-based stone;
[e] per ASTM C 241 or C 1353; f per ASTM C 880
[g] maximum acid resistance (in.) per ASTM C 217: 0.015 exterior; 0.025 interior
[h] first value is across grain; second value is along grain
[i] not recommended for paving in areas subject to heavy foot traffic

(Extracted, with permission, from ASTM C 503 *Standard Specification for Marble Dimension Stone (Exterior)*, ASTM C 568 *Standard Specification for Limestone Dimension Stone*, ASTM C 615 *Standard Specification for Granite Dimension Stone*, ASTM C 616 Quartz-Based Dimension Stone, and ASTM C 619 *Standard Specification for Slate Dimension Stone*, copyright ASTM International, 100 Barr Harbor Drive, West Conshohocken, PA 19428. Copies of the complete standards may be obtained from ASTM, phone: 610-832-9585, email: service@astm.org, website: www.astm.org).

cubes, square prisms, or cylinders. The minimum diameter or lateral dimension (distance between opposite faces) is 2 inches and the minimum ratio of height to diameter or lateral dimension is 1.1 inches. At least five specimens are required for each condition of testing. For granite, which requires testing wet and dry, and both perpendicular and parallel to the rift, twenty specimens are required.

In modulus of rupture testing of marble, limestone, granite, and quartz-based stone in accordance with ASTM C 99 and in flexure testing of slate in accordance with ASTM C 120, the application of load is along a single center line of the specimen span. Test specimens for ASTM C 99 are required to be 4 inches by 8 inches by 2.25 inches. Like compressive testing, five test specimens are required for each condition of testing. Test specimens for ASTM C 120 are required to be 1.5 inches wide by 12 inches long by 1.25 inches thick, and at least six specimens must be tested.

Flexural testing of marble in accordance with ASTM C 880 is performed on specimens that are 4 inches wide by 15 inches long by 1.25 inches thick. When tested, the specimens must have a span of at least 12.5 inches. At least five test specimens are required for each condition of test. This flexural testing consists of applying two lines of load, each at the quarter point of the span.

Evaluation of weather resistance of slate is performed in accordance with ASTM C 217. Three test specimens, each 2 inches by 4 inches, are required. Testing consists of determining the depth of softening in acid by one of two permitted methods: shear/scratch tester, which requires test equipment provided by the Taber Instrument Corporation; or by hand scraping tool. The specimens are submersed in 1 percent sulfuric acid solution for seven days, and the acid is replaced daily.

Evaluation of abrasion resistance may be performed either in accordance with ASTM C 241 or ASTM C 1353. ASTM C 241 requires three test specimens that are 2 inches square and 1 inch thick. This procedure utilizes a power-driven grinding lap that is revolved at 45 rpm. ASTM C 1353 requires three test specimens that are 4 inches square and ⅜ inch thick. This test method utilizes a Taber abraser, equipment that is manufactured by Taber Instrument Corporation.

The physical property requirements for the various types of dimension stone are listed in Table 1.25. Granite is the only type of dimension stone that is required to be tested in four conditions for compressive strength, flexural strength, and modulus of rupture: wet, dry, parallel to rift, and perpendicular to rift. The reported value is the minimum of the average strength of test specimens evaluated in the four conditions.

2

MORTAR AND GROUT

Mortar and grout are used to bind one masonry unit to another, and to bind together masonry units with ties, anchors, and reinforcement. The material ingredients of mortar and grout are similar to those used in producing concrete: cement, aggregate, and water. However, mortar and grout differ conceptually from concrete in these primary respects: water content; stiffness of the mix; aggregate size; and permitted cementitious materials.

Concrete is mixed with the minimum amount of water necessary to produce a batch that can be placed. Compressive strength is the most important property of concrete. Since increased water content reduces compressive strength, the ratio of water to cement is limited in concrete. The amount of water that is permitted to be added to field-mixed mortar and grout, however, is not limited. More water is needed in mortar and grout than in concrete to produce materials

fastfacts

Mortar and grout differ conceptually from concrete in these primary respects: water content; stiffness of the mix; aggregate size; and permitted cementitious materials.

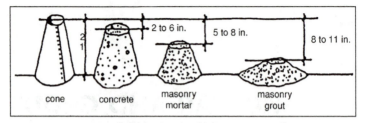

FIGURE 2.1 Relative Consistency of Concrete, Mortar, and Grout
(Beall & Jaffe, *Concrete and Masonry Databook,* McGraw-Hill, 2003)

with appropriate flow and the ability to hydrate the cement (permitting the mortar and grout to harden and gain strength) despite the suction action of the masonry units against which the mortar and grout are placed.

Concrete is generally mixed to a fairly stiff consistency, with slump between 2 and 6 inches. Mortar is mixed to a somewhat looser consistency to permit it to be spread on masonry units, but it must be stiff enough to support the weight of those units. Mortar slump is usually between 5 and 8 inches. Grout is mixed to a very fluid consistency, which permits it to be placed in and completely fill small spaces. Grout slump should be between 8 and 11 inches. These comparisons are shown in Figure 2.1.

The size of the aggregate used in mortar and grout is smaller than that used in concrete. Mortar aggregate is limited to sand. Grout aggregate may be sand alone or sand together with pea gravel.

Concrete, mortar, and grout are all permitted to have portland cement and blended cement as their cementitious materials. However, mortar may also incorporate lime, which is not used in concrete and may only be used sparingly in grout. Furthermore, mortar may be made with masonry cement or mortar cement, neither of which is used in grout or concrete.

fastfacts

Concrete, mortar, and grout are all permitted to have portland cement and blended cement as their cementitious materials. However, mortar may also incorporate lime, which is not used in concrete and may only be used sparingly in grout.

MORTAR AND GROUT COMPONENT MATERIALS

Both mortar and grout consist of cementitious materials and aggregate in combination with water. If the water used to mix mortar and grout is clean and drinkable, then it should be free of contaminants that might detract from the performance of those materials. Following is a more detailed discussion of the cementitious materials and aggregate used in mortar and grout.

Cementitious Materials

Table 2.1 lists the ASTM standards that govern the cements that may be used in mortar and grout. Portland cement (ASTM C 150), blended hydraulic cement (ASTM C 595), and hydraulic cement (ASTM C 1157) may each be used in mortar or grout. However, masonry cement (ASTM C 91) and mortar cement (ASTM C 1329) are only permitted in mortar and may not be used in grout. Each of these cement standards classifies the cement into a number of types. However, not all of the types are permitted to be used in mortar or grout. Table 2.2 lists cement types that are permitted in mortar and grout, and designates which may be used in mortar and which may be used in grout.

The most commonly referenced physical requirements for cements are listed in Table 2.3, including compressive strength, air content, and time of setting. Of the five standards that govern the various cements that can be used in mortar or grout, only ASTM C 1329 for mortar cement has a requirement for flexural bond strength. That requirement is for 70 psi for Type N mortar cement, 100 psi for Type S mortar cement, and 115 psi for Type M mortar cement. The test method for determining flexural bond strength is given in the Annex to ASTM C 1329. The standard warns that the flexural strength value determined by that test should not be regarded as the flexural strength capacity of a masonry wall built

TABLE 2.1   ASTM Standards for Cement in Mortar and Grout

ASTM Std. No.	Name ("Standard . . .)
C 91	Specification for Masonry Cement
C 150	Specification for Portland Cement
C 595	Specification for Blended Hydraulic Cement
C 1157	Performance Specification for Hydraulic Cement
C 1329	Specification for Mortar Cement

with the mortar cement. Variations occur between the test and actual performance of the construction because the test utilizes standardized concrete masonry units, whereas any type of masonry units may be used in the construction. ASTM C 1329 also warns that the flexural strength test should not be considered to

TABLE 2.2 Cement Types Permitted in Mortar and Grout

ASTM & Type	Definition	M[1]	G[2]
C 91 Masonry Cement			
M	for use in Type M mortar without further addition of cements or hydrated lime	yes	no
S	for use in Type S mortar without further addition of cements or hydrated lime	yes	no
N	for use in Type N mortar without further addition of cements or hydrated lime; and for use in Type M or Type S mortar when cements are added	yes	no
C 150 Portland Cement			
I	for use when the special properties for any other type are not required	yes	yes
IA	air-entraining cement for the same uses as Type I, where air-entrainment is desired	yes	yes
II	for general use, especially when moderate sulfate resistance or moderate heat of hydration is desired	yes	yes
IIA	air-entraining cement for the same uses as Type II, where air-entrainment is desired	yes	yes
III	for use when high early strength is desired	yes	yes
IIIA	air-entraining cement for the same uses as Type III, where air-entrainment is desired	yes	yes
C 595 Blended Hydraulic Cement			
IS	portland blast-furnace slag cement	yes	yes
IS(MS)	same as type IS, except with moderate sulfate resistance	no	yes
IS-A	same as type IS, except with air-entrainment	yes	yes
IS-A(MS)	same as type IS-A, except with moderate sulfate resistance	no	yes
IP	portland-pozzolan cement	yes	yes
IP-A	same as type IP, except with air-entrainment	yes	yes
I(PM)	pozzolan-modified portland cement	yes	no
I(PM)-A	same as type I(PM), except with air-entrainment	yes	no

[1] permitted in mortar per ASTM C 270
[2] permitted in grout per ASTM C 476

(continued)

TABLE 2.2 Cement Types Permitted in Mortar and Grout *(continued)*

ASTM & Type	Definition	M[1]	G[2]
C 1157 Hydraulic Cement			
GU	for general construction when one or more of the special types are not required	yes	yes
HE	high early-strength	yes	yes
MS	moderate sulfate resistance	yes	yes
HS	high sulfate resistance	yes	yes
MH	moderate heat of hydration	yes	no
LH	low heat of hydration	yes	no
C 1329 Mortar Cement			
M	for use in Type M mortar without further addition of cements or hydrated lime	yes	no
S	for use in Type S mortar without further addition of cements or hydrated lime	yes	no
N	for use in Type N mortar without further addition of cements or hydrated lime; and for use in Type M or Type S mortar when cements are added	yes	no

[1] permitted in mortar per ASTM C 270
[2] permitted in grout per ASTM C 476

(Extracted, with permission, from ASTM C 91 *Standard Specification for Masonry Cement,* ASTM C 150 *Standard Specification for Portland Cement,* ASTM C 595 Standard Specification for Blended Hydraulic Cement, ASTM C 1157 *Standard Performance Specification for Hydraulic Cement,* and ASTM C 1329 *Standard Specification for Mortar Cement,* copyright ASTM International, 100 Barr Harbor Drive, West Conshohocken, PA 19428. Copies of the complete standards may be obtained from ASTM, phone: 610-832-9585, email: service@astm.org, website: www.astm.org).

be representative of the extent of bond for the purposes of water permeance evaluation.

Other cementitious materials that may be used in mortar and grout are limes. Two ASTM standards govern the types of limes that may be used.

They are ASTM C 5 "Standard Specification for Quicklime for Structural Purposes" and ASTM C 207 "Standard Specification for Hydrated Lime for Masonry Purposes." Lime is used in combination with portland cement (ASTM C 150), blended hydraulic cement (ASTM C 595), or hydraulic cement (ASTM C 1157) but not with masonry cement (ASTM C 91) or mortar cement (ASTM C 1329).

Although hydrated lime can be combined directly with the other mortar ingredients, quicklime must first be slaked and aged. Slaking is the chemical reaction that produces hydrated lime when

TABLE 2.3 Physical Requirements for Cements in Mortar and Grout

ASTM and Type	ASTM C 266 Gilmore Time of Setting (min.)		Min. Compressive Strength[a] (psi) at		Air Content of Mortar[b] (% volume)	
	Initial Set, not less than	Final Set, not more than	7 Days	28 Days	Min.	Max.
ASTM C 91 Masonry Cement						
M	90	1440	1800	2900	8	19
S	90	1440	1300	2100	8	19
N	120	1440	500	900	8	21
ASTM C 150 Portland Cement						
I	60	600	2760	—	—	12
IA	60	600	2320	—	16	22
II	60	600	2470	1740	—	12
IIA	60	600	2030	1310	16	22
III	60	600	—	—	—	12
IIIA	60	600	—	—	16	22
ASTM C 1329 Mortar Cement						
M	90	1440	1800	2900	8	15
S	90	1440	1300	2100	8	15
N	120	1440	500	900	8	17

[a] per ASTM C 109 for portland cement; per modified ASTM C 109 for masonry cement and mortar cement.
[b] per ASTM C 91 for masonry cement; per ASTM C 185 for portland cement; per ASTM C 1329 for mortar cement.

(continued)

quicklime and water are mixed. After slaking, sand is added to the slaked quicklime and the mixture is stored for a minimum of 24 hours prior to use. ASTM C 207 defines four types of hydrated lime, which are listed in Table 2.4.

Aggregates

Sand that is used in mortar is governed by ASTM C 144 "Standard Specification for Aggregate for Masonry Mortar." Permitted sand may be natural or manufactured; manufactured sand is produced

fastfacts

Slaking is the chemical reaction that produces hydrated lime when quicklime and water are mixed.

TABLE 2.3 Physical Requirements for Cements in Mortar and Grout *(continued)*

ASTM and Type	ASTM C 191 Vicat Time of Setting (in.)		ASTM C 109 Min. Compressive Strength (psi) at		ASTM C 185 Air Content of Mortar (% volume)	
	Initial Set, not less than	Final Set, not more than	7 Days	28 Days	Min.	Max.
ASTM C 595 Blended Hydraulic Cement						
IS	45	420	2900	3620	—	12
IS(MS)	45	420	2610	3620	—	12
IS-A	45	420	2320	2900	16	22
IS-A(MS)	45	420	2030	2900	16	22
IP	45	420	2900	3620	—	12
IP-A	45	420	2320	2900	16	22
I(PM)	45	420	2900	3620	—	12
I(PM)-A	45	420	2320	2900	16	22
ASTM C 1157 Hydraulic Cement						
GU	45	420	—	4060	—	—
HE	45	420	—	—	—	—
MS	45	420	—	4060	—	—
HS	45	420	—	—	—	—
MH	45	420	—	3190	—	—
LH	45	420	—	—	—	—

(Extracted, with permission, from ASTM C 91 *Standard Specification for Masonry Cement,* ASTM C 150 Standard Specification for Portland Cement, ASTM C 595 *Standard Specification for Blended Hydraulic Cement,* ASTM C 1157 *Standard Performance Specification for Hydraulic Cement,* and ASTM C 1329 *Standard Specification for Mortar Cement,* copyright ASTM International, 100 Barr Harbor Drive, West Conshohocken, PA 19428. Copies of the complete standards may be obtained from ASTM, phone: 610-832-9585, email: service@astm.org, website: www.astm.org.

TABLE 2.4 ASTM C 207 Hydrated Lime Types

Type	Definition
N	normal hydrated lime for masonry purposes
S *	special hydrated lime for masonry purposes
NA	normal air-entraining hydrated lime for masonry purposes
SA *	special air-entraining hydrated lime for masonry purposes

* Type S and Type SA hydrated limes are differentiated from Type N and Type NA hydrated limes principally by their ability to develop high, early plasticity and higher water retentivity, and by a limitation on their unhydrated oxide content.

(Extracted, with permission, from ASTM C 207 *Standard Specification for Hydrated Lime for Masonry Purposes,* copyright ASTM International, 100 Barr Harbor Drive, West Conshohocken, PA 19428. A copy of the complete standard may be obtained from ASTM, phone: 610-832-9585, email: service@astm.org, website: www.astm.org).

fastfacts

The required flexural strength values in ASTM C 1329 for mortar cement are derived from testing that was performed on mortars made with portland cement and lime. That is, Type N portland cement-lime mortar was established as being able to achieve 70 psi flexural tensile strength, so Type N mortar cement is required to meet the same level of flexural tensile strength. The values for Type S mortar cement and Type M mortar cement have the same basis.

Mortars made with masonry cement and those made with air-entrained portland cement-lime generally do not achieve the same flexural strength as the equivalent types of mortar cement mortar or non-air-entrained portland cement-lime mortar. Building codes reduce the allowable flexural strength values of masonry built with masonry cement or air-entrained portland cement-lime mortar to 50 to 60 percent of the values given for masonry constructed with mortar cement mortar or non-air-entrained portland cement-lime mortar.

from crushed stone, gravel or air-cooled iron blast-furnace slag that is specifically processed to ensure suitable gradation. Shape of the sand particles is important, because excessive flat and elongated sand particles result in a mortar with reduced workability.

Gradation of the sand is also important. When the sand consists of particles of uniform size, whether that size is fine or coarse, the quantity of cementitious matrix required to coat each particle is greater than the quantity of cementitous matrix required to coat each particle of sand that is comprised of variable sized particles. This phenomenon occurs because uniformly sized particles do not fit together as tightly as variable sized particles, as illustrated in Figure 2.2.

The ASTM-required gradation of sand for use in masonry mortar is shown in Table 2.5. The standard explains that this gradation requirement may not be appropriate when the mortar joints exceed ½ inch in thickness. In that case, coarser aggregate conforming to the fine aggregate in ASTM C 404 "Standard Specification for Aggregate for Masonry Grout" is more suitable.

If the sand that is proposed to be used does not meet the gradation requirements shown in Table 2.5, it may still be used provided that the mortar so prepared complies with the property requirements of the ASTM standard for mortar. Those requirements

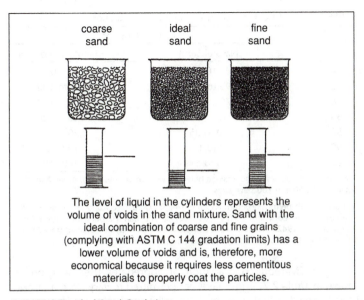

FIGURE 2.2   Ideal Sand Gradation
(Trowel Tips–Mortar Sand, Portland Cement Association)

are discussed later in this chapter. The amount of deleterious materials in samples of graded aggregate is limited by ASTM C 144, as shown in Table 2.6. The aggregate is further required to be free of injurious amounts of organic impurities.

Aggregates used in grout are governed by ASTM C 404 "Standard Specification for Aggregates for Masonry Grout." Permitted aggregates are natural or manufactured sand used alone or in combination with coarse aggregates. Manufactured sand and coarse aggregate are products obtained by crushing stone, gravel, or air-cooled iron blast-furnace slag.

fastfacts

Shape of the sand particles is important, because excessive flat and elongated sand particles result in a mortar with reduced workability.

TABLE 2.5 ASTM C 144 Masonry Mortar Sand Gradation

Sieve Size	Percent Passing[a]	
	Natural Sand	Manufactured Sand
4	100	100
8	95 to 100	95 to 100
16	70 to 100	70 to 100
30	40 to 75	40 to 75
50	10 to 35	20 to 40
100	2 to 15	10 to 25
200	0 to 5	0 to 10

[a] Not more than 50% shall be retained between any two sieve sizes nor more than 25% between No. 50 and 100 seive sizes.

(Extracted, with permission, from ASTM C 144 *Standard Specification for Aggregate for Masonry Mortar,* copyright ASTM International, 100 Barr Harbor Drive, West Conshohocken, PA 19428. A copy of the complete standard may be obtained from ASTM, phone: 610-832-9585, email: service@astm.org, website: www.astm.org).

The ASTM-required gradation of aggregates used in grout is shown in Table 2.7. Aggregates are permitted to vary from that gradation provided that they meet the following three requirements:

1. 100 percent of the fine aggregate passes the ⅜-inch sieve and no more than 5 percent of the natural sand (or 10 percent of manufactured sand) passes the No. 200 sieve;

2. 100 percent of the coarse aggregate passes the ½-inch sieve and no more than 5 percent passes the No. 30 sieve; and

3. grout compressive strength meets the requirements of ASTM C 476 "Standard Specification for Grout for Masonry."

The permitted limits of deleterious substances in the fine or coarse aggregate used in grout are the same as the requirements for mortar aggregate (see Table 2.6). Also similar to mortar aggregate, the fine aggregate must be free of injurious amounts of organic impurities.

fastfacts

Manufactured sand and coarse aggregate are products obtained by crushing stone, gravel, or air-cooled iron blast-furnace slag.

TABLE 2.6 ASTM C 144 Limitation of Deleterious Materials

Item	Max. Permitted Weight (%)
Friable particles	1.0
Lightweight particles, floating on liquid having specific gravity of 2.0	0.5[a]

[a] Does not apply to blast-furnace slag aggregate.

(Extracted, with permission, from ASTM C 144 *Standard Specification for Aggregate for Masonry Mortar,* copyright ASTM International, 100 Barr Harbor Drive, West Conshohocken, PA 19428. A copy of the complete standard may be obtained from ASTM, phone: 610-832-9585, email: service@astm.org, website: www.astm.org).

Admixtures

Admixtures are substances other than those specifically prescribed in the standard for masonry mortar (water, aggregate, and cementitious materials) that are used to improve one or more of the recognized desirable properties of conventional masonry mortar. Attributes of admixtures that are used in mortar are governed by ASTM C 1384 "Standard Specification for Admixtures for Masonry Mortars." As that standard explains, the acceptability of an admixture is dependent upon its performance in the admixed mortar. Acceptance of the admixed mortar is based on: attaining performance that is equivalent to that required for the conventional mortar or attaining improved performance of one or more properties while maintaining the required performance levels for other properties.

Five classifications of admixtures for mortar are defined by ASTM C 1384: bond enhancer, workability enhancer, set accelerator, set retarder, and water repellent. The advantages gained by the use of each of these admixtures are listed in Table 2.8. Physical requirements for these admixtures, shown in Table 2.9, are based on a comparison to the reference mortar. The reference mortar is mortar of the same composition as an admixed mortar, except that it does

fastfacts

Admixtures are substances other than those specifically prescribed in the standard for masonry mortar (water, aggregate, and cementitious materials) that are used to improve one or more of the recognized desirable properties of conventional masonry mortar.

TABLE 2.7 ASTM C 404 Masonry Grout Aggregate Gradation

Sieve Size	Amounts Finer Than Laboratory Sieve, Square Openings (weight %)				
	Fine Aggregate			Coarse Aggregae [b]	
	Size No. 1 [a]	Size No. 2 (ASTM C 144)		Size No. 8	Size No. 89
		Natural	Manufactured		
1/2 inch	—	—	—	100	100
3/8 inch	100	—	—	85 to 100	90 to 100
4	95 to 100	100	100	10 to 30	20 to 55
8	80 to 100	95 to 100	95 to 100	0 to 10	5 to 30
16	50 to 85	70 to 100	70 to 100	0 to 5	0 to 10
30	25 to 60	40 to 75	40 to 75	—	0 to 5
50	10 to 30	10 to 35	20 to 40	—	—
100	2 to 10	2 to 15	10 to 25	—	—
200	0 to 5	0 to 5	0 to 10	—	—

[a] Concrete sand specified in ASTM C 33 "Standard Specification for Concrete Aggregates".

[b] Standard sizes in ASTM D 448 "Standard Classification for Sizes of Aggregate for Road and Bridge Construction".

(Extracted, with permission, from ASTM C 404 *Standard Specification for Aggregates for Masonry Grout*, copyright ASTM International, 100 Barr Harbor Drive, West Conshohocken, PA 19428. A copy of the complete standard may be obtained from ASTM, phone: 610-832-9585, email: service@astm.org, website: www.astm.org).

TABLE 2.8 Masonry Mortar Admixture Classifications

Type	Improvement
bond enhancer	increases the bond strength between mortar and masonry unit
workability enhancer	increases the ease of working and using the mortar; increases board life and maintains water retention
set accelerator	shortens the setting time of a mortar
set retarder	lengthens the setting time of a mortar
water repellent	decreases the rate of water absorption of the mortar

(Extracted, with permission, from ASTM C 1384 *Standard Specification for Admixtures for Masonry Mortars,* copyright ASTM International, 100 Barr Harbor Drive, West Conshohocken, PA 19428. A copy of the complete standard may be obtained from ASTM, phone: 610-832-9585, email: service@astm.org, website: www.astm.org).

not include the admixture and may contain a different amount of water to obtain an equivalent flow or penetration as the admixed mortar. The physical requirements for admixed mortar, given in ASTM C 1384, are in addition to the physical requirements given in the ASTM specification standard (C 270) for mortar.

When specifically provided for in the project documents, calcium chloride may be used as a set accelerator, provided that the amount does not exceed 2 percent by weight of the portland cement con-

fastfacts

There is no acceptable admixture that is a true antifreeze. Some anti-freezes are actually misnomers for accelerators, but true antifreezes include alcohol. If the quantity of alcohol is sufficient to lower the freezing point of the mortar, compressive and bond strengths are reduced. Therefore, antifreeze admixtures are not recommended.

Set accelerator admixtures speed hydration of the portland cement. They do not prevent the mortar or grout from freezing. Calcium chloride is an effective accelerator, but it promotes corrosion of embedded metals in the masonry and can contribute to masonry efflorescence and spalling. Consequently, use of calcium chloride is not recommended. Set accelerators that conform to ASTM C 1384 should be used instead.

TABLE 2.9 ASTM C 1384 Physical Requirements for Admixed Masonry Mortar

Property	Bond Enhancer	Workability Enhancer	Set Accelerator	Set Retarder	Water Repellent
Minimum compressive strength (% of reference)					
7 day	80	80	80	70	80
28 day	80	80	80	80	80
Minimum water retention (% of reference)	report	100	report	report	report
Air content of plastic mortar	report	report	report	report	report
Minimum board life (% of reference)	report	120	report	120	report
Minimum flexural bond strength (% of reference)	110	—	—	—	—
Maximum rate of water absorption (% of reference, 24 hr.)	—	—	—	—	50
Time of setting (allowable deviation from reference, hr:min)					
Initial set: at least	—	—	1:00 earlier	1:00 later	—
not more than	1:00 earlier nor 1:30 later	1:00 earlier nor 1:30 later	3:30 earlier	8:00 later	1:00 earlier nor 1:30 later
Final set: at least	—	—	1:00 earlier	—	—
not more than	1:00 earlier nor 1:30 later	1:00 earlier nor 3:30 later	—	8:00 later	1:00 earlier nor 1:30 later

tent or 1 percent by weight of the masonry cement content, or both, of the mortar. However, calcium chloride should only be used with caution, since it may have a detrimental effect on metals and some wall finishes.

Admixtures to add color to mortar are governed by ASTM C 979 "Standard Specification for Pigments for Integrally Colored Concrete." Some of the requirements for these admixtures are similar to the requirements for ASTM C 1384 admixtures, while others are specific to the color change imparted by the pigment. Required properties of mortar pigments are listed in Table. 2.10, while typical pigments used to color mortar are shown in Table 2.11.

TABLE 2.10 Required Properties of ASTM C 979 Pigments

Property	Requirement (Test Method)
Water wettability	Readily mixes with water (ASTM C 979)
Alkalai resistance	No significant change in color when treated with sodium hydroxide (ASTM C 979)
Total sulfates	Shall not exceed 5.0 mass % of the original pigment sample (ASTM C 979)
Water solubility	Total soluble matter shall not exceed 2.0 mass % of original pigment sample (ASTM C 979)
Atmospheric curing stability	Magnitude of color difference between pigmented specimens cured in dry air and those cured at high relative humidity shall not be greater than the difference between two unpigmented specimens cured under the same conditions (ASTM C 979)
Light resistance	Exposed portions of the specimen shall show no significant differences in color from the unexposed portions (ASTM C 979)
Compressive strength	Not less than 90% of the control mixture (ASTM C 39)
Water cement ratio	Not greater than 110% of control mixture
Initial and final setting times	Not more than 1 hour earlier and not more than 1.5 hours later than the uncolored control specimen (ASTM C 403)
Air content	Not more than 1% greater nor less than 1% lower than the air content in the control specimen (ASTM C 173 or ASTM C 231)

(Extracted, with permission, from ASTM C 979 *Standard Specification for Pigments for Integrally Colored Concrete,* copyright ASTM International, 100 Barr Harbor Drive, West Conshohocken, PA 19428. A copy of the complete standard may be obtained from ASTM, phone: 610-832-9585, email: service@astm.org, website: www.astm.org).

TABLE 2.11   Coloring Pigments for Mortar

Mortar Color[1]	Pigments Used[2]
Black, Gray	Black iron oxide, mineral black, carbon black
Brown, Red	Red iron oxide, brown iron oxide, raw umber, burnt umber
Rose, Pink	Red iron oxide (varying amounts)
Buff, Cream, Ivory	Yellow ochre, yellow iron oxide
White	White cement and white sand (no pigments required)
Green	Chromium oxide, phthalocyanine green
Blue	Cobalt blue, ultramarine blue, phthalocyanine blue

[1] Color of finished mortar is affected by color of cement and aggregates.
[2] Synthetic iron oxides have more tinting power than natural iron oxides, so less pigment is required per unit of mortar to produce a given color. Synthetic oxides also produce brighter, cleaner colors.

(Beall & Jaffe, *Concrete and Masonry Databook*, McGraw-Hill, 2003).

MORTAR

Mortar that is used to construct masonry is governed by ASTM C 270 "Standard Specification for Mortar for Unit Masonry." That standard defines four types of mortar: N, S, M, and O. Of these, only Types N, S, and M are addressed in the MSJC Code and Specification. The International Building Code, however, includes Type O. An additional mortar type, Type K, is not used in modern construction, but may be used for repointing older buildings, as discussed in Chapter 6. Table 2.12 lists the appropriate uses for each mortar type, and Table 2.13 gives the recommended and alternate applications for each mortar type according to ASTM C 270.

ASTM C 270 allows mortar to be specified by either proportions or by properties. When specifying mortar by proportions, the relative volumetric quantities of the component materials are listed. The proportion specification is a prescriptive specification. When mortar

fastfacts

Calcium chloride should only be used with caution, since it may have a detrimental effect on metals and some wall finishes.

fastfacts

The letters used to designate mortar types were derived from the word "MASONWORK." Starting with the first letter of that word and deleting the second, fourth, sixth, and eighth letters leaves "M", "S", "N", "O", and "K". When listed in this sequence, the mortar types are ranked from highest compressive strength to lowest compressive strength.

is made by combining the specified materials in the specified amounts, the properties of that mortar can be predicted based on established historic usage. When specifying by properties, however, aspects of the mortar performance must be established by testing. The property specification is a performance specification. It is inappropriate to require that both the relative volumetric proportions be followed and that the performance requirements be met. Only one method should be used to specify mortar. Both methods yield mortar of equivalent quality.

When the project specifications do not detail whether the proportion specification method or the property specification method should be followed, the proportion specification is considered to govern. However, even under those circumstances, data can be presented to the specifier to demonstrate that a mortar that does not meet the proportion requirements still meets the property requirements.

TABLE 2.12 Mortar Types and Applications

Type	Application
N	for general use; best choice unless another type is needed for a specific purpose
S	alternative general use mortar with higher compressive strength; needed below grade and in high seismic regions
M	alternative mortar with highest compressive strength and lowest workability; alternative for below grade masonry and in high seismic regions
O	alternative mortar for non-bearing masonry and for repointing older masonry
K	alternative mortar for repointing historic masonry

TABLE 2.13 ASTM C 270 Guide for the Selection of Masonry Mortars

Location	Building Segment	Mortar Type Recom'd	Mortar Type Alt.
Exterior, above grade	Loadbearing walls	N	S or M
	Non-loadbearing walls	O[1]	N or S
	Parapet walls	N	S
Exterior, at or below grade	Foundation walls, retaining walls, sewers, manholes, walks[2], pavements[2], and patios[2]	S	M or N
Interior	Loadbearing walls	N	S or M
	Non-loadbearing partitions	O[1]	N

[1] Type O mortar is recommended for use where the masonry is unlikely to be frozen when saturated and unlikely to be subjected to high winds or other significant lateral loads. Type N or S should used in other cases.
[2] Masonry exposed to weather in a nominally horizontal surface is extremely vulnerable to weathering. Mortar for such masonry should be selected with due caution.

(Extracted, with permission, from ASTM C 270 *Standard Specification for Mortar for Unit Masonry,* copyright ASTM International, 100 Barr Harbor Drive, West Conshohocken, PA 19428. A copy of the complete standard may be obtained from ASTM, phone: 610-832-9585, email: service@astm.org, website: www.astm.org).

The ASTM C 270 proportion specification for mortar is shown in Table 2.14. These proportions are by volume, not weight. Table 2.15 provides bulk densities of mortar materials so that volume proportions can be converted into weights for batching. Note that two air-entraining materials are not permitted to be combined in the same mortar

fastfacts

Bond strength between mortar and masonry units is usually more important to the performance of the masonry assembly than compressive strength. Mortars with high compressive strength are usually produced at the expense of workability, flow, and water retentivity, which are properties that are necessary to achieve the maximum possible extent of bond (for water penetration resistance) and bond strength (for flexural tensile capacity). Therefore, the mortar type with the lowest compressive strength that is appropriate to the masonry's location and application should be utilized.

TABLE 2.14 ASTM C 270 Mortar Proportions by Volume

Cementitious Materials	Type	Portland Cement or Blended Cement	Masonry Cement Type M	Masonry Cement Type S	Masonry Cement Type N	Mortar Cement Type M	Mortar Cement Type S	Mortar Cement Type N	Hydrated Lime or Lime Putty	Aggregate (Sand), Measured in a Damp, Loose Condition
Cement-Lime	M	1	—	—	—	—	—	—	¼	not less than 2¼ and not more than 3 times the sum of the separate volumes of cement and lime
	S	1	—	—	—	—	—	—	over ¼ to ½	
	N	1	—	—	—	—	—	—	over ½ to 1¼	
	O	1	—	—	—	—	—	—	over 1¼ to 2½	
Mortar Cement	M	1	—	—	—	—	—	1	—	
	M	—	—	—	—	1	—	—	—	
	S	½	—	—	—	—	—	1	—	
	S	—	—	—	—	—	1	—	—	
	N	—	—	—	—	—	—	1	—	
	O	—	—	—	—	—	—	1	—	
Masonry Cement	M	1	—	—	1	—	—	—	—	
	M	—	1	—	—	—	—	—	—	
	S	½	—	—	1	—	—	—	—	
	S	—	—	1	—	—	—	—	—	
	N	—	—	—	1	—	—	—	—	
	O	—	—	—	1	—	—	—	—	

fastfacts

When mortar is made by combining the specified materials in the specified amounts, the properties of that mortar can be predicted based on established historic usage.

when using the proportion method for specifying mortar. Refer to Table 2.2 for types of cement that are permitted to be used in mortar.

When using Table 2.14 to make mortar, care should be taken to note the distinction between the way that the cementitous materials are specified versus the way that the aggregates are specified. The cement and lime volumes are stated in relative proportions; that is, for each unit volume of cement (one part), the appropriate unit volume of lime is zero when the cement is masonry cement or mortar cement and is between 0.25 and 2.5 depending upon the mortar type when the cement is portland cement or blended cement. The appropriate volume of aggregate, however, is based on the sum of the volumes of cement and lime (both are cementitous materials) and not on the unit volume of cement alone. The permitted aggregate ratio is given as a range, rather than a single number, to account for the variation in workability that results from subtle differences in particle gradation and shape.

The ASTM C 270 physical requirements for mortar that is specified by properties is shown in Table 2.16. When a mortar is specified by properties, a *laboratory-prepared* sample of the proposed mortar

TABLE 2.15   Bulk Densities of Mortar Materials

Material	Bulk Density
portland cement	94 pcf
blended cement	85 to 94 pcf *
mortar cement	*
masonry cement	70 to 90 pcf *
lime putty	80 pcf
hydrated lime	40 pcf
sand	80 pcf

* See weight on bag or obtain from supplier.

(Panaraese, Kosmatka, & Randall, *Concrete Masonry Handbook*, Portland Cement Association, 1991).

TABLE 2.16 ASTM C 270 Mortar Property Requirements[a, b]

Cementitious Materials	Type	Min. Average Compressive Strength at 28 Days (psi)	Minimum Water Retention (%)	Maximum Air Content[c] (%)
Cement-Lime	M	2500	75	12
	S	1800	75	12
	N	750	75	14
	O	350	75	14
Mortar Cement	M	2500	75	12
	S	1800	75	12
	N	750	75	14
	O	350	75	14
Masonry Cement	M	2500	75	18
	S	1800	75	18
	N	750	75	20
	O	350	75	20

[a] Applicable to laboratory-prepared mortar only.

[b] The aggregate ratio, measured in a damp, loose condition, shall not be less than 2-1/4 and not more than 3 times the sum of the separate volumes of cement and lime.

[c] When structural reinforcement is incorporated in cement-lime or mortar cement mortars, maximum air content shall not exceed 12%. When structural reinforcement is incorporated in masonry cement mortars, maximum air content shall not exceed 18%. Air content of non-air-entrained portland cement-lime mortar is generally less than 8 %.

(Extracted, with permission, from ASTM C 270 *Standard Specification for Mortar for Unit Masonry*, copyright ASTM International, 100 Barr Harbor Drive, West Conshohocken, PA 19428. A copy of the complete standard may be obtained from ASTM, phone: 610-832-9585, email: service@astm.org, website: www.astm.org).

fastfacts

The appropriate volume of aggregate, however, is based on the sum of the volumes of cement and lime (both are cementitous materials) and not on the unit volume of cement alone.

mix must be tested for compliance with these properties. The test methods are provided in ASTM C 270. When a mortar mix has thus been established to be in compliance with the standard's physical properties, the relative proportions of that mix are then used to make mortar in the field—with one exception. The exception is water content.

Water in the laboratory-prepared mortar mix that is tested for physical property compliance is limited to that which will produce a flow of 110 plus or minus 5 percent. This water content is not sufficient to produce a mortar with workable consistency suitable for laying masonry units in the field. Mortar used in the field must be mixed with the maximum amount of water that is consistent with workability so that the initial rate of absorption of the masonry units (suction) can be satisfied. That is, the mortar must be able to accommodate the masonry units that will reduce the mortar water content by absorbing some water (and cement fines) into the unit pores. The reduced water content of the laboratory-prepared mix mimics the water content of the field-mixed mortar after some water has been absorbed by the masonry units against which it has been placed. The properties of field-prepared mortar, with its higher water content, will differ from those listed in Table 2.16. Therefore, field-prepared mortar cannot be compared to the requirements of ASTM C 270 for evaluation of compliance.

ASTM C 780 "Standard Test Method for Preconstruction and Construction Evaluation of Mortars for Plain and Reinforced Unit Masonry" may be used to evaluate plastic and hardened mortar, either before or during its use in construction. ASTM C 1324 "Standard Test Method for Examination and Analysis of Hardened Masonry Mortar" may be used to determine the proportions of component materials in hardened mortars after placement. There is no ASTM test method to determine whether a hardened mortar sample complies with the property specifications of ASTM C 270.

fastfacts

When a mortar is specified by properties, a laboratory-prepared sample of the proposed mortar mix must be tested for compliance with these properties.

ASTM C 780 describes seven tests that may be performed to evaluate mortar. A brief summary of the purpose of each test follows.

Consistency determination by cone penetration: Prior to construction, this test is used to gauge the water addition for the various mortar mixes being contemplated for use. During construction, this test may be used to as a quick reference for indicating batch-to-batch variation in mix ingredients and mixing time. Note however, that erratic consistency readings indicate poor control during batching and mixing but do not indicate whether the cement, sand, or water additions were improper.

Consistency retention by cone penetration: Prior to construction, this test establishes the early-age setting and stiffening characteristics of the mortar mixes being contemplated for use. Because this test is performed in the laboratory, it allows comparison of performance of the various mortar systems. In the field, however, varying weather conditions may change the performance relationships.

Water content determination: This test measures the total water content of the mortar. The preconstruction test value serves as the control or basis for testing that is performed during construction; the test indicates the ability of the mixing operator to properly and consistently add water to the mixer.

fastfacts

Mortar used in the field must be mixed with the maximum amount of water that is consistent with workability so that the initial rate of absorption of the masonry units (suction) can be satisfied.

Aggregate ratio testing: Prior to construction, this test determines the ratio of cementitious material to aggregate when used in conjunction with the water content determination test. The preconstruction test value serves as the control for testing during construction; repeated testing shows the ability of the mixing operator to properly and consistently add cementitious materials and sand to the mortar mixer and establishes batch-to-batch variations in the mortar composition.

Air content testing: Establishing the air content of mortars that contain air-entraining cementitious materials or admixtures is important. Repeated testing during construction shows the changes that result from variations in mixing time, mixing efficiency, and other job-site conditions.

Compressive strength testing: Compression testing of molded mortar cylinders or cubes permits establishment of the strength developing characteristics of the mortar. Measured strength will vary with the mortar water content at the time of set. The measured values do not represent the actual strength of the mortar in the constructed masonry, because some water is suctioned out of the mortar when placed against the masonry units. Strength results will also vary with the shape of the specimen that is tested; cylinder compressive strength is considered equivalent to 85 percent of the cube compressive strength.

Splitting tensile strength: Measured splitting tensile strength will also vary with the mortar water content at the time of set. Similar to compressive strength, the splitting tensile strength measured by this test approximates, but does not represent, the actual strength of the in-place mortar.

ASTM C 1324 defines how to analyze hardened mortar by a combination of petrographic (microscopic) examination and chemical analysis to determine the component materials and their relative proportions. Thus, the mortar type, as defined in ASTM C 270 (M, S, N, or O), can be identified. Procedures included in the test method require a substantial degree of both petrographic and chemical skills as well as relatively elaborate equipment. If a mortar that is analyzed by ASTM C 1324 is determined to have a composition that does not conform to the proportion requirements of ASTM C 270 (Table 2.14), one cannot conclude that the mortar does not meet the requirements of the standard. The mortar may satisfy the alternate property requirements (Table 2.16) of ASTM C 270.

fastfacts

Fine grout is made with fine aggregates (sand) only. Coarse grout is made with both fine aggregates (sand) and coarse aggregates (pea gravel).

GROUT

Grout used in masonry construction is defined by ASTM C 476 "Standard Specification for Grout for Masonry." Two grout types are classified by that standard: fine and coarse. Fine grout is made with fine aggregates (sand) only. Coarse grout is made with both fine aggregates (sand) and coarse aggregates (pea gravel). The appropriate type of grout to be used in the construction is dependent upon the size of the grout space to be filled and the height to which the masonry is constructed prior to placing the grout. Grout placement is discussed in Chapter 6.

Similar to the approach taken by ASTM C 270 for mortar, grout may be specified by either a proportion method or by strength requirements. Testing is not required for the proportion or prescriptive approach but is needed for the strength or performance approach. It is inappropriate to require that both the relative volumetric proportions be followed and that the strength requirements be met. Only one method should be used to specify grout. Both methods yield grout of equivalent quality.

The proportion requirements for grout are given in Table 2.17. These proportions are based on relative volumes of materials, not weights. Refer to Table 2.15 for bulk densities of materials used in mortar and grout so that volumetric proportions can be converted into weight proportions. Refer to Table 2.2 for the types of cements that may be used in grout. Note that masonry cement

fastfacts

Note that masonry cement and mortar cement are not permitted in grout, and the use of lime in grout is entirely optional.

TABLE 2.17 ASTM C 476 Grout Proportions by Volume

Grout Type	Parts by Volume of Portland Cement or Blended Cement	Parts by Volume of Hydrated Lime or Lime Putty	Aggregate, Measured in a Damp, Loose Condition	
			Fine	Coarse
Fine	1	0 to 1/10	2-1/4 to 3 times the volumes of the cement and lime	—
Coarse	1	0 to 1/10	2-1/4 to 3 times the volumes of the cement and lime	1 to 2 times the sum of the volumes of the cement and lime

fastfacts

Masonry surfaces that will be in contact with the grout are lined with a thin permeable material to prevent bonding to the masonry units.

and mortar cement are not permitted in grout, and the use of lime in grout is entirely optional. When lime is added, it is only used in small quantities.

When using Table 2.17 to make grout, care should be taken to note the distinction between the way that the cementitious materials are specified versus the way that the aggregates are specified. The cement and lime volumes are stated in relative proportions; that is, for each unit volume of cement (one part), the appropriate unit volume of lime is between zero and 1/10 (0 to 1/10 part). The appropriate volume of aggregate, however, is based on the sum of the volumes of cement and lime (both are cementitious materials) and not on the unit volume of cement alone. The permitted aggregate ratio is given as a range, rather than a single number, to account for the variation in workability that results from subtle differences in particle gradation and shape.

When grout is alternatively specified by strength, both the compressive strength and the slump must be evaluated. The strength requirements for grout are given in Table 2.18.

Compressive strength testing of grout specimens is performed in accordance with ASTM C 1019 "Standard Test Method for Sampling and Testing Grout." This test may be used either to help select grout proportions by comparing test values or as a quality control test for uniformity of grout properties during construction. ASTM C 1019 defines the grout specimen as having a square cross-section whose

TABLE 2.18 ASTM C 476 Grout Strength Requirements

Property	Requirement	Test Method
compressive strength at 28 days	2,000 psi	ASTM C 1019
slump	8 in. to 11 in.	ASTM C 143

(Extracted, with permission, from ASTM C 476 *Standard Specification for Grout for Masonry,* copyright ASTM International, 100 Barr Harbor Drive, West Conshohocken, PA 19428. A copy of the complete standard may be obtained from ASTM, phone: 610-832-9585, email: service@astm.org, website: www.astm.org).

sides are a minimum of 3 inches and whose height is twice the side dimension. The permitted dimensional tolerance is 5 percent. A test set consists of at least three specimens that are tested at each age at which the strength value must be evaluated.

Grout specimens are formed by placing grout between the same types of masonry units that are used in the construction. If the grout is to be placed between different types of units, both types are used to form the grout mold. The units are placed in a pinwheel pattern, illustrated in Figure 2.3, arranged so as to form the required size grout specimen. A non-absorbent block is used at the bottom of the grout space if needed to adjust the grout specimen height to achieve the required aspect ratio. Masonry surfaces that will be in contact with the grout are lined with a thin permeable material to prevent bonding to the masonry units. Other methods of forming grout specimens, such as drilling grout-filled cores of regular masonry units or filling compartments in slotted corrugated cardboard boxes specifically manufactured for that purpose, are not explicitly permitted by ASTM C 476. Because test results may vary with these methods, comparative test results between the specimen formed as described in the standard and

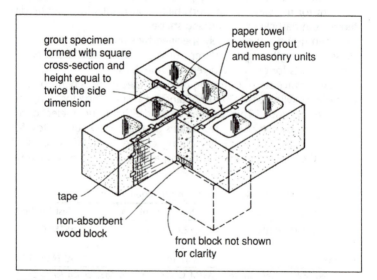

FIGURE 2.3 Grout Specimen Formed with Concrete Masonry Units
(Panarese, Kosmatka, & Randall, *Concrete Masonry Handbook,* Portland Cement Association, 1991)

the proposed method of forming the specimen should be performed to establish the relationship.

ASTM C 1019 discusses how to fill the grout mold as well as how to store, transport, cure, and test the grout specimens. The mold should be removed between 24 and 48 hours after making the specimens. Within 72 hours after initial molding, the specimens should be stored in a special moist room. The specimens are capped prior to testing, and test results are reported to the nearest 10 psi. The amount of grout slump, illustrated in Figure 2.1, must also be reported together with the compressive strength.

chapter 3

REINFORCEMENT

Steel reinforcement is used to enhance the strength and ductility of masonry. It increases both axial and flexural compressive and tensile resistance. It also makes the masonry able to deform a greater amount prior to breaking, giving greater warning of failure.

Steel reinforcement used in masonry may be deformed bars, placed either vertically or horizontally in masonry cells or in the collar joint between masonry wythes, welded wire fabric placed in the collar joint between masonry wythes, or joint reinforcement that is placed horizontally in the bed joints between masonry courses. Placement of steel reinforcement is discussed in Chapter 7. This chapter discusses the materials used to make steel reinforcement and their governing ASTM standards.

STEEL REINFORCING BARS

The ASTM standards that apply to steel reinforcing bars are listed in Table 3.1. ASTM A 615, A 706, and A 996 each describe the type of steel used to make the reinforcement. ASTM A 767 and A 775 define the corrosion protection coating that may be applied to any of the three types of steel reinforcing bars.

Steel reinforcing bars are made in sizes No. 3 through No. 18 for concrete reinforcement. However, only sizes No. 3 through No. 11 are permitted to be used in masonry. The bar designations are derived from the nominal diameter of the bar in ⅛-inch increments;

TABLE 3.1   ASTM Standards for Steel Reinforcing Bars

ASTM	Name ("Standard Specification for . . .")
A 615	Deformed and Plain Billet-Steel Bars for Concrete Reinforcement
A 706	Low-Alloy Steel Deformed and Plain Bars for Concrete Reinforcement
A 996	Rail-Steel and Axle-Steel Deformed Bars for Concrete Reinforcement
A 767	Zinc-Coated (Galvanized) Steel Bars for Concrete Reinforcement
A 775	Epoxy-Coated Steel Reinforcing Bars

a No. 5 bar has a nominal diameter of ⅝ inch, and so on. Because the bar is deformed, the actual overall diameter of the bar is slightly larger than its nominal designation. Typical bar deformations are illustrated in Figure 3.1 and approximate overall bar diameters are listed in Table 3.2.

All three standards for steel reinforcing bars (A 615, A 706, and A 996) provide the same dimensional and deformation requirements for steel bars. These requirements are shown in Table 3.3.

The three reinforcing bar standards classify the bars by grade, where the grade designation is based on the minimum yield strength of the bar. For example, grade 50 bars have a yield strength of 50,000 pounds per square inch (psi). Minimum tensile strength of the bar is higher than its yield strength. ASTM A 996 also classifies the bars by type: type A, type R, and type "rail symbol." Type A bars may be grade 40 or grade 50, whereas type R and type "rail symbol" bars may be grade 50 or grade 60.

Reinforcing bars are fabricated with a standardized marking system to identify the mill that produced the bar, the bar size, the steel type, and the yield of the steel. Variations in the marking systems are illustrated in Figure 3.2.

fastfacts

Two standards that formerly applied to steel reinforcing bars were ASTM A 616 "Standard Specification for Rail-Steel Deformed and Plain Bars for Concrete Reinforcement" and ASTM A 617 "Standard Specification for Axle-Steel Deformed and Plain Bars for Concrete Reinforcement." These standards were withdrawn in September 1999 and replaced by ASTM A 996.

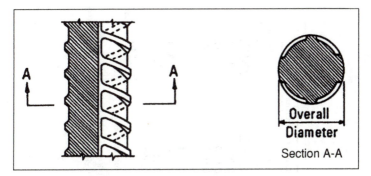

FIGURE 3.1 Deformations on Steel Reinforcing Bars
(*Manual of Standard Practice,* Concrete Reinforcing Steel Institute, 1973).

There are some differences between the three standards in terms of the bar sizes and grades that are included. One significant difference is that ASTM A 706 is the only standard of the three that limits the chemical composition and carbon equivalent of the steel that is used to make the bar so that weldability is enhanced. The physical requirements for steel reinforcing bars are shown in Table 3.4.

When reinforcing bars must be spliced, several alternative methods are available. When the steel material permits it (as in ASTM A 706), the bar ends may be welded. Welded connections must be capable of resisting 125 percent of the bar's specified yield strength. Bar ends may also be overlapped, either in contact or not in contact. Requirements for lap splicing are discussed in Chapter 7. Alternatively, bar ends may be connected by a mechanical coupler, as illustrated in Figure 3.3. Mechanical couplers must be capable of resisting 125 percent of the bar's specified yield strength.

TABLE 3.2 Overall Diameters of Steel Reinforcing Bars

Bar Designation No.	Approximate Diameter to Outside of Deformations (in.)
3	7/16
4	9/16
5	11/16
6	7/8
7	1
8	1-1/8
9	1-1/4
10	1-7/16
11	1-5/8

(*Manual of Standard Practice,* Concrete Reinforcing Steel Institute, 1973)

TABLE 3.3   Section Properties of Deformed Steel Reinforcing Bars

| Bar No. | Nominal Dimensions* | | | | Deformation Requirements | | |
	Diameter (in.)	Cross-Sectional Area (sq.in.)	Perimeter (in.)	Nominal Weight (plf)	Maximum Average Spacing (in.)	Minimum Average Height (in.)	Max. Gap (Chord of 12.5% of Nominal Perimeter)
3	0.375	0.11	1.178	0.376	0.262	0.015	0.143
4	0.500	0.20	1.571	0.668	0.350	0.020	0.191
5	0.625	0.31	1.963	1.043	0.437	0.028	0.239
6	0.750	0.44	2.356	1.502	0.525	0.038	0.286
7	0.875	0.60	2.749	2.044	0.612	0.044	0.334
8	1.000	0.79	3.142	2.670	0.700	0.050	0.383
9	1.128	1.00	3.544	3.400	0.790	0.056	0.431
10	1.270	1.27	3.990	4.303	0.889	0.064	0.487
11	1.410	1.56	4.430	5.313	0.987	0.071	0.540

* Nominal dimensions of deformed bars are the same as a round bars having the same weight per foot as the deformed bar.

(Extracted, with permission, from ASTM A 615 *Standard Specification for Deformed and Plain Billet-Steel Bars for Concrete Reinforcement*, ASTM A 706 *Standard Specification for Low-Alloy Steel Deformed and Plain Bars for Concrete Reinforcement*, and ASTM A 996 *Standard Specification for Rail-Steel and Axle-Steel Deformed Bars for Concrete*.

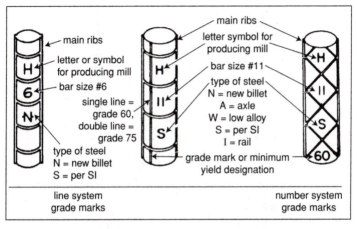

FIGURE 3.2 Identifying Designations on Steel Reinforcing Bars
(Beall & Jaffe, *Concrete and Masonry Databook,* McGraw-Hill, 2003).

fastfacts

Reinforcing bars are fabricated with a standardized marking system to identify the mill that produced the bar, the bar size, the steel type, and the yield of the steel.

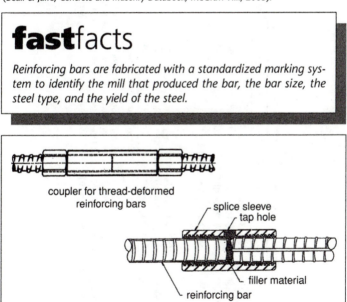

FIGURE 3.3 Reinforcing Bar Coupler
(ACI 439.3R-91, *Mechanical Connections of Reinforcing Bars,* American Concrete Institute International).

TABLE 3.4 Physical Requirements for Steel Reinforcing Bars

ASTM	Bar Sizes	Grade	Min. Yield Strength (psi)	Min. Tensile Strength (psi)
A615	No. 3 through No. 6	40	40,000	60,000
Billet Steel	No. 3 through No. 11, 14 and 18	60	60,000	90,000
	No. 6 through 11, 14 and 18	75	75,000	100,000
A706	No. 3—11, 14 and 18	60	60,000	80,000
Low-Alloy Steel*				
A 996	No. 3 through No. 8	40	40,000	70,000
Rail-Steel and Axle-Steel	No. 3 through No. 8	50	50,000	80,000
	No. 3 through No. 8	60	60,000	90,000

* ASTM A706 requires a maximum yield strength of 78,000 psi and tensile strength at least 1.25 times the actual yield strength.

(Extracted, with permission, from ASTM A 615 *Standard Specification for Deformed and Plain Billet-Steel Bars for Concrete Reinforcement*, ASTM A 706 *Standard Specification for Low-Alloy Steel Deformed and Plain Bars for Concrete Reinforcement*, and ASTM A 996 *Standard Specification for Rail-Steel and Axle-Steel Deformed Bars for Concrete Reinforcement*, copyright ASTM International, 100 Barr Harbor Drive, West Conshohocken, PA 19428. Copies of the complete standards may be obtained from ASTM, phone: 610-832-9585, email: service@astm.org, website: www.astm.org).

TABLE 3.5 ASTM A 767 Zinc Coating Requirements

Class	Bar Size	Minimum Mass (Weight) of Zinc Coating (oz/sq. ft. of surface)
I	No. 3	3.00
	No. 4 and larger	3.50
II	No. 3 and larger	2.00

(Extracted, with permission, from ASTM A 767 *Standard Specification for Zinc-Coated (Galvanized) Steel Bars for Concrete Reinforcement*, copyright ASTM International, 100 Barr Harbor Drive, West Conshohocken, PA 19428. A copy of the complete standard may be obtained from ASTM, phone: 610-832-9585, email: service@astm.org, website: www.astm.org).

Corrosion protection for steel reinforcing bars is provided by masonry cover over the steel. Additional corrosion protection may be provided by coating the bars with either zinc or epoxy. ASTM A 767 defines two classes of zinc coating, where Class I requires a thicker application of zinc than Class II, as shown in Table 3.5. Coating damage that occurs during fabrication or handling to the point of shipment is required to be repaired with a zinc-rich formulation in accordance with ASTM A 780 "Standard Practice for Repair of Hot Dip Galvanized Coatings." Sheared ends of reinforcing bars are required to be repaired in the same manner.

ASTM A 775 defines only one category of epoxy coating, but provides several performance requirements for that coating. The requirements for the epoxy coating on steel reinforcing bars are shown in Table 3.6.

TABLE 3.6 ASTM A 775 Epoxy Coating Requirements

Property	Requirement
thickness	at least 90% of measurements shall be 7 to 12 mils; reject if more than 5% of measurements are less than 5 mils
continuity	no more than one (1) holiday per foot
flexibility	no cracking or disbanding of coating on the outside radius of the bar bent in accordance with ASTM A 775
adhesion	average coating disbondment radius shall not exceed 0.16 in. when measured from the edge of the intentional coating defect and tested in accordance with ASTM G 8

(Extracted, with permission, from ASTM A 775 *Standard Specification for Epoxy-Coated Steel Reinforcing Bars*, copyright ASTM International, 100 Barr Harbor Drive, West Conshohocken, PA 19428. A copy of the complete standard may be obtained from ASTM, phone: 610-832-9585, email: service@astm.org, website: www.astm.org).

Welded wire fabric is composed of cold-drawn steel wire, as-drawn or galvanized, that is fabricated into sheets or rolls by electric resistance welding.

WELDED WIRE FABRIC

Welded wire fabric is composed of cold-drawn steel wire, as-drawn or galvanized, that is fabricated into sheets or rolls by electric resistance welding. The fabric consists of a series of longitudinal and

TABLE 3.7   ASTM A 82 Dimensional Requirements for Plain Wire

Size	Nominal Diameter, (in.)	Nominal Area, (in.)
W 0.5	0.080	0.005
W 1.2	0.124	0.012
W 1.4	0.134	0.014
W 2	0.160	0.020
W 2.5	0.178	0.025
W 2.9	0.192	0.029
W 3.5	0.211	0.035
W 4	0.226	0.040
W 4.5	0.239	0.045
W 5	0.252	0.050
W 5.5	0.265	0.055
W 6	0.276	0.060
W 8	0.319	0.080
W 10	0.357	0.100
W 12	0.391	0.120
W 14	0.422	0.140
W 16	0.451	0.160
W 18	0.479	0.180
W 20	0.505	0.200
W 22	0.529	0.220
W 24	0.533	0.240
W 26	0.575	0.260
W 28	0.597	0.280
W 30	0.618	0.300
W 31	0.628	0.310
W 45	0.757	0.450

(Extracted, with permission, from ASTM A 82 *Standard Specification for Steel Wire, Plain, for Concrete Reinforcement,* copyright ASTM International, 100 Barr Harbor Drive, West Conshohocken, PA 19428. A copy of the complete standard may be obtained from ASTM, phone: 610-832-9585, email: service@astm.org, website: www.astm.org).

transverse wires arranged substantially at right angles to each other and welded together at points of intersection. The wires used to make the fabric may be plain or deformed.

Welded wire fabric is governed by one of two ASTM standards:

- A 185 "Standard Specification for Steel Welded Wire Fabric, Plain, for Concrete Reinforcement"
- A 497 "Standard Specification for Steel Welded Wire Fabric, Deformed, for Concrete Reinforcement"

TABLE 3.8 ASTM A 496 Dimensional Requirements for Deformed Wire

Size	Nominal Diameter, (in.)	Nominal Area, (in.)
D-1	0.113	0.01
D-2	0.159	0.02
D-3	0.195	0.03
D-4	0.225	0.04
D-5	0.252	0.05
D-6	0.276	0.06
D-7	0.299	0.07
D-8	0.319	0.08
D-9	0.338	0.09
D-10	0.356	0.10
D-11	0.374	0.11
D-12	0.390	0.12
D-13	0.406	0.13
D-14	0.422	0.14
D-15	0.437	0.15
D-16	0.451	0.16
D-17	0.465	0.17
D-18	0.478	0.18
D-19	0.491	0.19
D-20	0.504	0.20
D-21	0.517	0.21
D-22	0.529	0.22
D-23	0.541	0.23
D-24	0.553	0.24
D-25	0.564	0.25
D-26	0.575	0.26
D-27	0.586	0.27
D-28	0.597	0.28
D-29	0.608	0.29
D-30	0.618	0.30
D-31	0.628	0.31
D-45	0.757	0.45

(Extracted, with permission, from ASTM A 496 *Standard Specification for Steel Wire, Deformed, for Concrete Reinforcement,* copyright ASTM International, 100 Barr Harbor Drive, West Conshohocken, PA 19428. A copy of the complete standard may be obtained from ASTM, phone: 610-832-9585, email: service@astm.org, website: www.astm.org).

The wires that make up ASTM A 185 welded wire fabric must conform to ASTM A 82 "Standard Specification for Steel Wire, Plain, for Concrete Reinforcement." The "plain" term in the title indicates that the wire is not deformed. Wires that make up ASTM A 497 welded wire fabric may be either plain (ASTM A 82) or deformed per ASTM A 496 "Standard Specification for Steel Wire, Deformed, for Concrete Reinforcement."

Wire sizes are designated by a letter and a number. The letter denotes whether the wire is plain (letter W) or deformed (letter D). The number indicates the nominal cross-sectional area in one-hundredths of a square inch. Dimensional requirements for plain wire are shown in Table 3.7, and the dimensional requirements for deformed wire are given in Table 3.8. The physical requirements for both plain and deformed wires are shown in Table 3.9.

Welded wire fabric comes in rolls or sheets. The ordering length and width of sheets is presented in Figure 3.4. Commonly used sizes of welded wire fabric are listed in Table 3.10, which also lists the previous designation format for welded wire fabric.

When the wires in the welded wire fabric are to be galvanized for corrosion protection, the zinc coating is required to comply with the requirements for regular coating, as specified in ASTM A 641 "Standard Specification for Zinc-Coated (Galvanized) Carbon Steel Wire." According to ASTM A 641, wires produced with a regular coating are required to have the full surface covered with zinc, but there is no specified minimum weight of coating.

TABLE 3.9 Physical Requirements for Plain and Deformed Wires for Welded Wire Fabric

Wire	Min. Tensile Strength (psi)	Minimum Yield Strength (psi)
A 82 Plain, size W 1.2 and larger	75,000	70,000
A 82 Plain, smaller than size W 1.2	70,000	56,000
A 496 Deformed	80,000	70,000

(Extracted, with permission, from ASTM A 82 *Standard Specification for Steel Wire, Plain, for Concrete Reinforcement,* and ASTM A 496 *Standard Specification for Steel Wire, Deformed, for Concrete Reinforcement,* copyright ASTM International, 100 Barr Harbor Drive, West Conshohocken, PA 19428. Copies of the complete standards may be obtained from ASTM, phone: 610-832-9585, email: service@astm.org, website: www.astm.org).

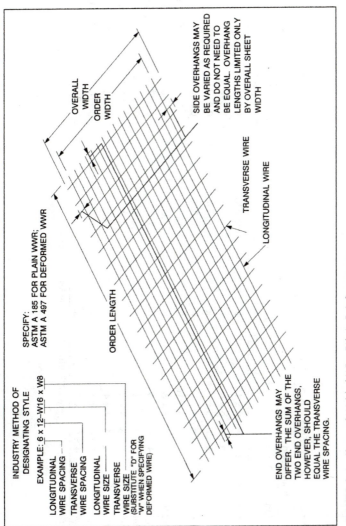

FIGURE 3.4 Sheet of Welded Wire Fabric Reinforcement
(*PCI Design Handbook*, 5th ed., Precast/Prestressed Concrete Institute, 1999).

TABLE 3.10   Common Welded Wire Fabric Sizes and Properties

Style Designation		Roll (R) or Sheet (S)	Longitudinal Steel Area (sq.in/ft)	Transverse Steel Area (sq.in/ft)	Approximate Total Weight (lb/100 sq.ft)
Current Designation (by W-number)	Previous Designation (by steel wire gauge)				
6 x 6—W1.4 x W1.4	6 x 6—10 x 10	R	0.028	0.028	21
6 x 6—W2.0 x W2.0	6 x 6—8 x 8	R	0.040	0.040	30
6 x 6—W2.9 x W2.9	6 x 6—6 x 6	R	0.058	0.058	42
6 x 6—W4.0 x W4.0	6 x 6—4 x 4	R	0.080	0.080	58
4 x 4—W1.4 x W1.4	4 x 4—10 x 10	R	0.042	0.042	31
4 x 4—W2.0 x W2.0	4 x 4—8 x 8	R	0.060	0.060	44
4 x 4—W2.9 x W2.9	4 x 4—6 x 6	R	0.087	0.087	62
4 x 4—W4.0 x W4.0	4 x 4—4 x 4	R	0.120	0.120	85
6 x 6—W2.9 x W2.9	6 x 6—6 x 6	S	0.058	0.058	42
6 x 6—W4.0 x W4.0	6 x 6—4 x 4	S	0.080	0.080	58
6 x 6—W5.5 x W5.5	6 x 6—2 x 2	S	0.110	0.110	80
4 x 6—W4.0 x W4.0	4 x 6—4 x 4	S	0.120	0.080	85

(Adapted from *PCI Design Handbook*, 5th ed., Precast/Prestressed Concrete Institute, 1999).

fastfacts

Joint reinforcement consists of two to four longitudinal wires with cross wires that are welded to the longitudinal wires with or without fixed tabs or adjustable ties. The cross wires may be perpendicular to the longitudinal wires (ladder type) or may be diagonal to those wires (truss type).

JOINT REINFORCEMENT

Masonry joint reinforcement, which is governed by ASTM A 951 "Standard Specification for Masonry Joint Reinforcement," is made in three primary forms: rigid ladder or truss type, fixed tab-type, and adjustable ties. These forms are illustrated in Figures 3.5 through 3.8. Joint reinforcement consists of two to four longitudinal wires with cross wires that are welded to the longitudinal wires with or without fixed tabs or adjustable ties. The cross wires may be perpendicular to the longitudinal wires (ladder type) or may be diagonal to those wires (truss type).

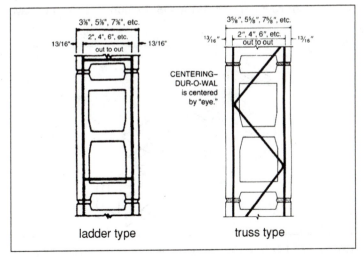

FIGURE 3.5 Two-Wire Joint Reinforcement (Dur-O-Wal, Inc., Aurora, IL).

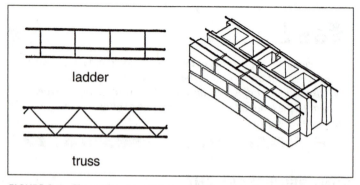

FIGURE 3.6   Three-Wire Joint Reinforcement
(Beall & Jaffe, *Concrete and Masonry Databook,* McGraw-Hill, 2003).

Joint reinforcement is typically fabricated in 10-foot lengths. To accommodate wall intersections, joint reinforcement is commonly prefabricated in T-shapes and L-shapes, as shown in Figure 3.9.

The wires in joint reinforcement can either be plain wire conforming to ASTM A 82 or stainless steel wire conforming to ASTM A 580, Type 304 ("Standard Specification for Stainless and Heat-Resisting Steel Wire"). The longitudinal wires in joint reinforcement must be deformed. Deformations are made by knurling as defined in ASTM A 951. Therefore, the deformed wires in joint reinforcement do not conform to ASTM A 496. Deformed wires must have

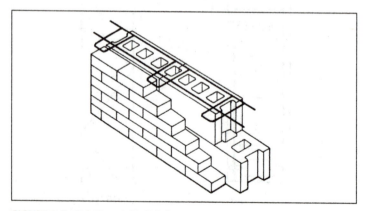

FIGURE 3.7   Tab-Type Joint Reinforcement
(Panarese, Kosmatka, & Randall, *Concrete Masonry Handbook,* Portland Cement Association, 1991).

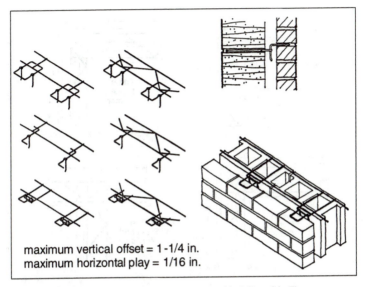

maximum vertical offset = 1-1/4 in.
maximum horizontal play = 1/16 in.

FIGURE 3.8 Two-Wire Joint Reinforcement with Adjustable Ties
(Beall & Jaffe, *Concrete and Masonry Databook,* McGraw-Hill, 2003).

one set of two deformations, spaced at 0.7 times the wire diameter, with at least eight sets per inch.

The minimum size of wire for use in joint reinforcement is W1.1 (11 gauge). Wire sizes that are used in joint reinforcement are shown in Table 3.11. Joint reinforcement with longitudinal wires of size W 1.7 (9 gauge) is considered standard. When the longitudinal wires are size W 2.1 (8 gauge), the joint reinforcement is medium, and when the longitudinal wires are size W 2.8 (3/16 inch), the joint reinforcement is heavy-duty.

TABLE 3.11 Dimensional Properties of Wires in Masonry Joint Reinforcement

ASTM Wire Size	Wire Gauge No.	Nominal Diameter (in.)	Nominal Area (sq.in.)	Nominal Perimeter (in.)
W1.1	11	0.1205	0.0114	0.379
W1.7	9	0.1483	0.0173	0.466
W2.1	8	0.1620	0.0206	0.509
W2.8	3/16 in.	0.1875	0.0276	0.589
W4.9	1/4 in.	0.2500	0.0491	0.785

(Beall & Jaffe, *Concrete and Masonry Databook,* McGraw-Hill, 2003)

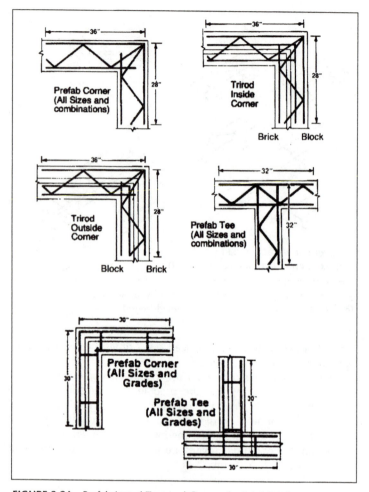

FIGURE 3.9A Prefabricated Tees and Corners for Joint Reinforcement (Dur-O-Wal, Inc., Aurora, IL).

The physical requirements for the wire used in joint reinforcement, as given in ASTM A 951, are shown in Table 3.12. Since joint reinforcement may be used as structural reinforcing in masonry, the area of steel reinforcement provided is shown in Table 3.13.

When a corrosion-protection coating is provided for the plain steel wires, the coating must either conform to ASTM A 641 "Standard Specification for Zinc-Coated (Galvanized) Carbon Steel Wire" (mill

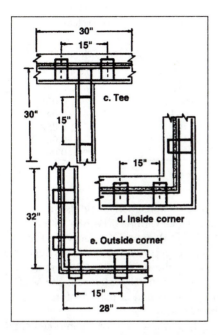

galvanized) or to ASTM A 153 "Standard Specification for Zinc Coating (Hot-Dip) on Iron and Steel Hardware," Class B. The mill galvanized coating is required to have a minimum thickness of 0.1 ounces per square foot and may be applied prior to fabrication of the joint reinforcement. The hot-dip galvanized coating is required to have a minimum thickness of 1.5 ounces per square foot and must be applied after fabrication.

TABLE 3.12   ASTM A 951 Physical Requirements for Wires in Joint Reinforcement

Property	Requirement
Tensile strength	80,000 psi
Yield strength	70,000 psi
Minimum reduction of area (ruptured section from tensile test)	30 %
Weld shear strength	25,000 times the nominal area of the larger wire

(Extracted, with permission, from ASTM A 951 *Standard Specification for Masonry Joint Reinforcement,* copyright ASTM International, 100 Barr Harbor Drive, West Conshohocken, PA 19428. A copy of the complete standard may be obtained from ASTM, phone: 610-832-9585, email: service@astm.org, website: www.astm.org).

TABLE 3.13 Steel Reinforcement Area (in.^{2}/ft.) Provided by Joint Reinforcement

Size and Number of Side Rods	Vertical Spacing of Joint Reinforcement	
	8 in. on center	16 in. on center
2–W 1.7 (9 ga.)	0.052	0.026
2–W 2.1 (8 ga.)	0.061	0.031
2–W 2.8 (3/16 in.)	0.083	0.041
4–W 1.7 (9 ga.)	0.104	0.052
4–W 2.1 (8 ga.)	0.122	0.061
4–W 2.8 (3/16 in.)	0.166	0.083

(Dur-O-Wal, Aurora, IL)

Corrosion protection for joint reinforcement may also be provided by epoxy coating in accordance with ASTM A 884 "Standard Specification for Epoxy-Coated Steel Wire and Welded Wire Fabric for Reinforcement," Class B, Type 2, according to the MSJC Specification. To satisfy this requirement, the epoxy coating must be at least 18 mils thick. IBC 2000, which generally references the MSJC documents, does not permit epoxy coating of joint reinforcement, however, and requires it to be galvanized.

Mill-galvanized coating is only permitted for joint reinforcement that is placed in interior walls. When the joint reinforcement will be placed in exterior walls or in interior walls exposed to a mean relative humidity in excess of 75 percent (such as natatoria), the coating must be hot-dip galvanizing (or epoxy unless IBC is the governing building code).

fastfacts

Mill-galvanized coating is only permitted for joint reinforcement that is placed in interior walls.

chapter 4

THE MASONRY ASSEMBLY

Masonry assemblies are generally categorized as either single wythe or multiple wythe. Single wythe assemblies include walls that are one unit thick. Single wythe walls may be used to resist vertical loads, out-of-plane lateral loads, and/or in-plane lateral loads. They may be single-span, multiple-span, or cantilevers. Single wythe masonry members are connected to lateral support elements, such as floor or roof diaphragms or columns, only at the edges of the wall section.

Multiple wythe masonry assemblies are subdivided by the Masonry Standards Joint Committee (MSJC) Code into structurally composite systems and structurally non-composite systems. The composite systems are required to have either masonry headers to connect the wythes or a completely filled collar joint between the wythes and regularly spaced metal ties that connect the wythes. The types of masonry units need not be the same in each wythe of a multi-wythe composite masonry assembly. A structurally composite masonry system is also known as a functional barrier. This terminology relates to the way in which the system prevents water from passing through it and entering the enclosed space. In a composite or barrier system, water that enters the exterior masonry wythe is stopped at the collar joint (barrier) and prevented from traveling inward beyond that point. Examples of structurally composite masonry walls are shown in Figure 4.1.

Collar joints in structurally non-composite systems are required to be free of masonry headers, grout, and mortar. They may contain

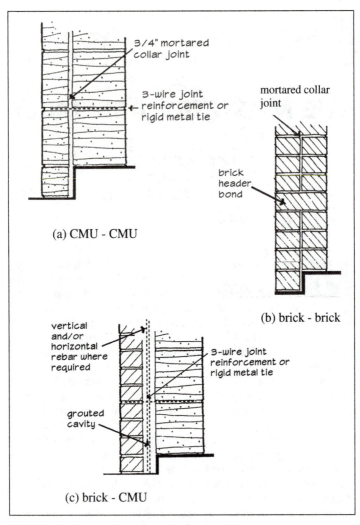

FIGURE 4.1   Structurally Composite (Barrier) Walls
(Beall & Jaffe, *Concrete and Masonry Databook,* McGraw-Hill, 2003).

rigid insulation or drainage mat material. The individual wythes in a structurally non-composite masonry system are required to be connected by regularly spaced metal ties that span across the open collar joint. Like a composite system, the types of masonry units need not be the same in each wythe of a multi-wythe non-composite masonry assembly. A structurally non-composite masonry system is also called

fastfacts

A structurally composite masonry system is also known as a functional barrier. This terminology relates to the way in which the system prevents water from passing through it and entering the enclosed space.

a cavity wall or a drainage wall. This terminology derives from the way this type of system prevents water entry to the enclosed space. In a drainage wall, water that enters the exterior masonry wythe is permitted to run (drain) down the back face of that wythe (via the open collar joint or cavity) until it reaches a through-wall flashing system that collects the water and directs it to the exterior. Structurally non-composite walls are illustrated in Figure 4.2.

Muliple wythe walls may be single-span, multiple-span, or cantilevers. They may be used to resist vertical loads, out-of-plane lateral loads, and/or in-plane lateral loads. In a composite masonry system, all wythes act together to resist these loads. In a non-composite masonry system, only the wythe that is directly loaded acts to resist that load, except for out-of-plane lateral loads, which are distributed to each wythe in proportion to that wythe's relative stiffness.

Veneers are a special subcategory of multi-wythe non-composite walls. Veneers are considered to be an architectural facing element and are treated as if they have no capacity to resist loads other than their own self-weight. They rely on their backing material, which may be masonry, concrete, steel studs, or wood studs, to resist all applied loads. Therefore, regularly spaced anchors throughout the field of the veneer are required to connect the veneer to its backing and to transfer applied loads. Like non-composite walls, the veneer and its backing system provide an internal drainage plan to handle water penetration.

fastfacts

Veneers are considered to be an architectural facing element and are treated as if they have no capacity to resist loads other than their own self-weight.

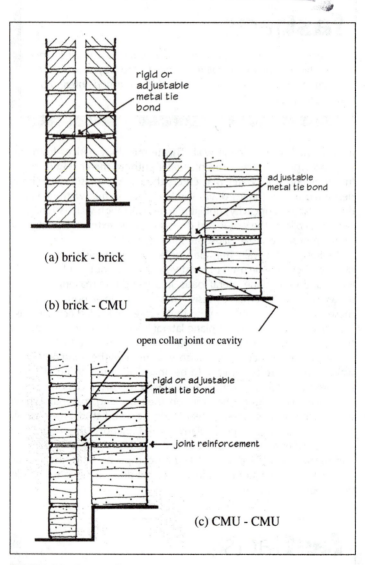

(a) brick - brick

(b) brick - CMU

(c) CMU - CMU

rigid or
adjustable
metal tie
bond

adjustable
metal tie bond

open collar joint or cavity

rigid or adjustable
metal tie bond

joint reinforcement

FIGURE 4.2 Structurally Non-Composite (Cavity) Walls
(Beall & Jaffe, *Concrete and Masonry Databook,* McGraw-Hill, 2003).

fastfacts

A structurally non-composite masonry system is also called a cavity wall or a drainage wall. This terminology derives from the way this type of system prevents water entry to the enclosed space.

The masonry assembly, which consists of the combination of masonry units, mortar, grout (if used), and reinforcement (if used), may be tested for several structural properties. Standard test methods exist to evaluate masonry compressive strength, bond strength, and shear strength.

Appearance of the assembly, relative to unit chippage, distortion, and dimensional variations, is addressed by the ASTM standards for the masonry units rather than by a standard for the assembly. However, a standard is available to evaluate serviceability of the masonry assembly relative to water penetration.

The test methods for evaluating the masonry assembly are presented in this chapter.

COMPRESSIVE STRENGTH

According to the International Building Code (IBC), and to the Masonry Standards Joint Committee (MSJC), author of "Building Code Requirements for Masonry Structures" (ACI 530/ASCE 5/TMS 402), "Specification for Masonry Structures" (ACI 530.1/ASCE 6/TMS 602), compressive strength of a masonry assembly can be determined by one of two methods: unit strength method or prism test method.

The unit strength method is simple and does not require testing. This method can be used provided that the following three criteria are met:

- Masonry units conform to their relevant ASTM standard.
- Mortar bed joint thickness does not exceed ⅝ inch.
- In grouted masonry, the grout either conforms to ASTM C 476 or the minimum grout compressive strength equals f'_m, the specified compressive strength of the masonry assembly, but is not less than 2000 psi.

TABLE 4.1   IBC 2000 Compressive Strength of Clay Masonry

Required Net Area Compressive Strength of Clay Masonry Units (psi)		For Net Area Compressive Strength of Masonry (psi)
When used with Type M or S Mortar	When used with Type N Mortar	
2,400	3,000	1,000
4.400	5,500	1,500
6,400	8,000	2,000
8,400	10,500	2,500
10,400	13,000	3,000
12,400	——	3,500
14,400	——	4,000

(International Building Code 2000).

When the materials and construction conform to these requirements, the compressive strength of the masonry assembly can be determined by referencing tables. In each table, the compressive strength of the masonry assembly is based upon the mortar type and the masonry unit compressive strength. Table 4.1 provides the compressive strength of clay masonry according to IBC 2000. Table 4.2 provides the compressive strength of clay masonry according to IBC 2003 and Table 4.3 gives the compressive strength of concrete masonry according to IBC.

When the criteria for using the unit strength method are not met or the specified compressive strength of masonry exceeds the values listed in the tables, the prism strength method must be

TABLE 4.2   IBC 2003 Compressive Strength of Clay Masonry

Required Net Area Compressive Strength of Clay Masonry Units (psi)		For Net Area Compressive Strength of Masonry (psi)
When used with Type M or S Mortar	When used with Type N Mortar	
1,700	2,100	1,000
3,350	4,150	1,500
4,950	6,200	2,000
6,600	8,250	2,500
8,250	10,300	3,000
9,900	——	3,500
13,200	——	4,000

(International Building Code 2003, MSJC Specification 1999, and MSJC Specification 2002).

TABLE 4.3   IBC Compressive Strength of Concrete Masonry

Required Net Area Compressive Strength of Concrete Masonry Units (psi)		For Net Area Compressive Strength of Masonry (psi)
When used with Type M or S Mortar	When used with Type N Mortar	
1,250	1,300	1,000
1,900	2,150	1,500
2,800	3,050	2,000
3,750	4,050	2,500
4,800	5,250	3,000

(International Building Code 2000, IBC 2003, MSJC Specification 1999, and MSJC Specification 2002).

used to evaluate masonry assembly compressive strength. ASTM C 1314 "Standard Test Method for Compressive Strength of Masonry Prisms" describes the procedure for evaluating the compressive strength of masonry assemblies. Test values so obtained are compared to the specified compressive strength of masonry, f'_m, required by the project specifications in accordance with the relevant building code.

The compressive strength of veneer assemblies and glass unit masonry assemblies need not be determined because their structural design does not depend upon this parameter. Single wythe and multiple wythe masonry assemblies that are designed by the empirical or "rule-of-thumb" provisions of the applicable building code are also exempted from the requirement for determining assembly compressive strength for the same reason.

fastfacts

In IBC, tabulated values of masonry compressive strength based on the unit strength method are based on the values provided by MSJC. The f'_m values listed in MSJC changed in the 1999 edition of that standard, when the reference test method for prism testing changed from ASTM C 447 to ASTM C 1314. However, the values shown in the 2000 edition of IBC did not change from the earlier (1995) edition of MSJC and, therefore, appear to be in error. The IBC tabulated values were corrected to agree with MSJC in IBC 2003.

Prisms for testing of compressive strength are always constructed in sets of three, as required by ASTM C 1314. A prism set is required for each combination of materials and each test age at which the compressive strength of the masonry is to be determined. The testing laboratory that performs the test is required to be certified in accordance with ASTM C 1093 "Standard Practice for Accreditation of Testing Agencies for Unit Masonry."

Prisms must be made of masonry units that are representative of those that will be used in the actual construction, with one exception. For units that have flutes or ribs that project at least ½ inch, the flutes or ribs must be removed by saw cutting flush with the unit at the base of the flute or rib. When the construction includes multi-wythe masonry in which the wythes have different units or mortar types, prism sets are required for each different wythe for separate testing.

Prisms are constructed by laying the units in stack bond in stretcher position, as shown in Figure 4.3. The full length of the units may be used to form the prism, or the units may be reduced in length to no less than 4 inches. Mortar material and joint thickness used to make the prism should correspond to those in the actual construction, but full mortar bedding should be used even if face-shell bedding will be used in the project. Mortar bedding is discussed further in Chapter 6. Prisms are built to a minimum of two units high, with a height to thickness ratio between 1.3 and 5.0. Prisms are built inside a moisture-tight bag, which is sealed upon completion of prism construction and maintained until two days prior to testing. If the unit cells or collar joint will be grouted in the construction, grouted prisms should be fabricated.

Newly fabricated prisms must be stored in a way to prevent them from freezing. They are kept undisturbed for 48 hours after construction or grouting, whereupon they are transported to the laboratory for storage at room temperature. To prevent damage during handling and transportation, the prisms must strapped or clamped and secured.

fastfacts

Prisms must be made of masonry units that are representative of those that will be used in the actual construction.

In the laboratory, the prisms are capped on the top and bottom with sulfur-filled capping or high-strength gypsum cement to produce smooth, level, and parallel surfaces. The test machine is required to have an accuracy of plus or minus 1.0 percent over the anticipated load range. The laboratory's test report will include the failure load for each prism, the compressive strength of each prism, and the average compressive strength of the prism set.

Prism compressive strength is calculated as the failure load divided by the net cross-sectional area for hollow unit prisms and as the failure load divided by the gross cross-sectional area for solid unit prisms and for grouted prisms. Values are reported to the nearest 10 psi. To account for variations in the height-to-thickness ratios

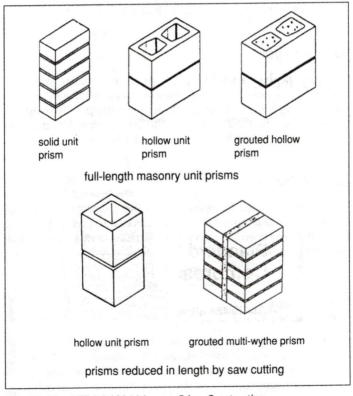

solid unit prism

hollow unit prism

grouted hollow prism

full-length masonry unit prisms

hollow unit prism   grouted multi-wythe prism

prisms reduced in length by saw cutting

FIGURE 4.3 ASTM C 1314 Masonry Prism Construction
(Panarese, Kosmatka, & Randall, *Concrete Masonry Handbook,* Portland Cement Association, 1991).

fastfacts

Prisms are built inside a moisture-tight bag, which is sealed upon completion of prism construction and maintained until two days prior to testing. If the unit cells or collar joint will be grouted in the construction, grouted prisms should be fabricated.

of prisms, ASTM C 1314 provides correction factors that must be applied to the calculated prism compressive strength to derive the masonry compressive strength. As shown in Table 4.4, the correction factor for a prism with height to thickness ratio of 2 is 1.0, meaning that this aspect ratio is the baseline.

Bond Strength

Four test methods are provided by ASTM to evaluate the bond strength of masonry assemblies. They are listed in Table 4.5. Differences among these standards include the method of forming the test specimen, method of loading the test specimen, and overall purpose. Each test method is summarized in this section.

ASTM C 952

ASTM C 952 is used for research into the bonding properties of masonry units and mortar and is not intended to predict the bond strength of commercial masonry construction. It provides for two test procedures: tensile bond strength of clay brick units using

fastfacts

Prism compressive strength is calculated as the failure load divided by the net cross-sectional area for hollow unit prisms and as the failure load divided by the gross cross-sectional area for solid unit prisms and for grouted prisms. Values are reported to the nearest 10 psi.

TABLE 4.4 ASTM C 1314 Correction Factors for Masonry Prism Compressive Strength

h_p/t_p*	Correction Factor
1.3	0.75
1.5	0.86
2.0	1.0
2.5	1.04
3.0	1.07
4.0	1.15
5.0	1.22

* h_p/t_p is the ratio of prism height to least lateral dimension of the prism.

(Extracted, with permission, from ASTM C 1314 *Standard Test Method for Compressive Strength of Masonry Prisms,* copyright ASTM International, 100 Barr Harbor Drive, West Conshohocken, PA 19428. A copy of the complete standard may be obtained from ASTM, phone: 610-832-9585, email: service@astm.org, website: www.astm.org).

crossed-brick couplets and flexural bond strength of concrete masonry units using two-unit prisms.

The crossed-brick couplet testing can be used to evaluate the bond between a specified brick and a specified mortar; the relative bond between different bricks and a specified mortar, or the relative bond between a specified brick and different mortars. Specimens are constructed in the laboratory, using a jig to ensure consistent placement of one brick perpendicularly to the one below with a mortar bed between the two. Completed specimens are kept covered for seven days, then stored at a minimum of 50 percent relative humidity. Specimens are tested in a machine that pulls the brick units apart, thus measuring tensile bond strength.

The concrete block prism is formed by laying two units with face shell mortar bedding in the manner specified by ASTM C 952. The concrete masonry unit must be preconditioned to an equivalent rel-

TABLE 4.5 ASTM Test Methods for Evaluating Bond Strength of Masonry Assemblies

ASTM Std. No.	Title ("Standard Test Method for . . .")
C 952	Bond Strength of Mortar to Masonry Units
C 1072	Measurement of Masonry Flexural Bond Strength
C 1357	Evaluating Masonry Bond Strength
E 518	Flexural Bond Strength of Masonry

fastfacts

Prisms cut from existing masonry must be handled so as to prevent the joints from becoming subject to tensile stresses.

ative humidity of not less than 50 percent. Completed specimens are kept covered for seven days, then stored at a minimum of 50 percent relative humidity. Specimens are tested in a compression machine that applies load eccentrically. Test results are used to compute flexural bond strength.

ASTM warns that variability is a characteristic of masonry bonding. ASTM C 952 advises that coefficients of variation of 15 to 35 percent are commonplace.

ASTM C 1072

ASTM C 1072 determines comparative values of flexural bond strength of non-reinforced masonry. The procedure involves physical testing of each mortar joint in the masonry test specimen.

Bond strengths determined from this test method can be used to evaluate the compatibility of mortars and masonry units; or to determine the effect of various factors on flexural bond strength, including masonry unit or mortar properties, workmanship, curing conditions, and coatings on masonry units. Flexural bond strength determined by this test method should not be interpreted as the flexural bond strength of the constructed masonry wall, but can be used to predict that strength. Also, results do not indicate the extent of bond for purposes of water permeance evaluation.

Test specimens for ASTM C 1072 may be fabricated in the laboratory, constructed in the field, or cut from existing masonry construction. Test specimens consist of prisms, two or more units in height, with a minimum width of 4 inches. When possible, the width should be one full masonry unit. A test set includes a minimum of five joint tests. Laboratory-prepared specimens should be cured in laboratory air (75 plus or minus 15 degrees F, relative humidity between 30 and 70 percent), and field-prepared specimens should be cured at atmospheric conditions similar to those in the masonry structure that they are intended to represent. Prisms cut from existing masonry must be handled so as to prevent the joints from becoming subject to tensile stresses.

Prisms are tested in a bond wrench frame that twists one of the masonry units off the remainder of the prism. The direction of loading is perpendicular to the length of the masonry unit. Flexural tensile strength is reported to the nearest psi and is based on gross area for solid units and on net area for hollow units. The report should include a description of the failure mode; that is, whether the failure occurred at the top of the mortar joint, at the bottom of the mortar joint, or at both locations.

ASTM C 1357

ASTM C 1357 is used to evaluate masonry flexural bond strength normal to bed joints. Two different test methods are included, one for laboratory-prepared specimens and one for field-prepared specimens. Both methods use the test procedure described by ASTM C 1072.

The purpose of the "Test Method for Laboratory-Prepared Specimens" is to compare bond strengths of various mortars or of one mortar under various conditions. Prisms are constructed using "standard" concrete masonry units rather than conventional masonry units. The "standard" concrete masonry units are 100 percent solid bricks, manufactured with the materials, proportions, curing conditions, and moisture content specified in the Annex to ASTM C 1357. Mortars are batched by weight equivalents of volume proportions and are mixed to a prescribed flow. Prisms are constructed using a jig for consistency and are bag-cured. A set of test specimens consists of no fewer than 15 nor more than 30 mortar joints. Because this test method uses controlled conditions of fabrication and curing that are not intended to represent field conditions, results should not be interpreted as the flexural bond strength of a wall constructed under field conditions using conventional masonry units.

The purpose of the "Test Method for Field-Prepared Specimens" is to evaluate the bond strength of a particular mortar-unit combi-

fastfacts

The purpose of the ASTM C 1357 "Test Method for Field-Prepared Specimens" is to evaluate the bond strength of a particular mortar-unit combination, either for pre-construction evaluation of materials or for quality control purposes during construction.

nation, either for pre-construction evaluation of materials or for quality control purposes during construction. Mortars are batched conventionally and flow is not prescribed. Prisms are constructed conventionally, without a jig, and are bag-cured. A test set consists of any convenient number of prisms containing a total of at least 15 mortar joints. The number of prisms should be limited to the number that can be constructed within 30 minutes. When performing field tests for quality control, one set of test prisms should be constructed for each 5000 square feet of masonry in the structure. Reported results include individual bond strength for each joint of the test specimens, the mean, the standard deviation, and the test age for each set of test specimens.

Many criteria are available for evaluating test results. Examples include requiring the minimum value to be not less than some target value or requiring the average value to be not less than a target value. However, these simple criteria are insensitive to the scatter of the test results. Because test results with smaller scatter are more reliable, the effect of scatter should be considered. The Appendix to ASTM C 1357 suggests two possible criteria for evaluating the results obtained. The criteria are illustrative only and are not mandatory.

The first suggested criterion compares the lower 10% characteristic value (10% fractile) with a minimum target value. This approach is appropriate when so many joints are tested (at least 30) that the sample closely approximates the entire population of joints.

The second suggested criterion is more complex than the characteristic value approach. The objective of this approach is to establish a 90% probability (confidence) that 90% of the masonry in the entire population will have flexural bond strength that equals or exceeds the target value. This statistical technique is referred to as a one-side tolerance limit. This approach is appropriate when relatively few joints are tested and thus the sample may not closely approximate the entire population of joints. It is also appropriate when the scatter of test results is small.

ASTM E 518

The test methods described in ASTM E 518 provide a simplified and economical means for gathering comparative research data on the flexural bond strength developed with different types of masonry units and mortar. They can also be used to monitor job quality control. These test methods are not intended to establish design strength of the masonry assembly; ASTM E 72 "Standard Methods of Conducting Strength Tests of Panels for Building Construction" should be used for that purpose.

Two test methods are provided in ASTM E 518. Test Method A utilizes a simply supported beam with third-point loading, and Test Method B utilizes a simple supported beam with uniform loading applied by an air bag. A test set consists of a minimum of five test specimens constructed as stack bond prisms at least 18 inches high. Masonry units and mortar representative of the construction are used to fabricate the specimens. A jig is used to maintain alignment of the prism and units are laid in the mortar bedding (full or face shell) that is specified for the project.

Prisms are cured for 28 days under laboratory conditions (75 plus or minus 15 degrees F and 30 to 70 percent relative humidity). When the testing is performed for quality control, the prisms should be cured in conditions similar to those in the masonry structure.

Prisms are laid horizontally on supports as a simply supported beam. Loading is then applied at a uniform rate. Test results are reported as the individual and average modulus of rupture (to the nearest psi), based either on gross area for solid masonry units or on net bedded area for hollow units. The standard deviation and the coefficient of variation are also reported.

WATER PERMEANCE

ASTM E 514 "Standard Test Method for Water Penetration and Leakage Through Masonry" is used to evaluate resistance to water penetration and leakage through unit masonry assemblages that are subject to wind-driven rain. It can aid in evaluating the effect of quality of materials, coatings, wall design, and workmanship. This test method is primarily used to establish

fastfacts

Water permeance is closely related to extent of mortar-to-unit bond (as opposed to bond strength), since most water penetrates the masonry assembly at the joints, where mortar and units interface. Compatible materials, good workmanship, and appropriate precautions during weather extremes lead to optimal extent of bond.

comparative behavior between various masonry wall construc-
tions in a given laboratory.

The testing apparatus used in the ASTM E 514 procedure, shown
in Figure 4.4, is a chamber that encloses a minimum of 12 square
feet. The perimeter of the chamber, which is placed in contact with
the masonry specimen, is lined with a closed cell compressible gasket material that can provide a water-tight seal. The face of the
chamber includes an observation port. Water is supplied to the
chamber from the top and into a spray pipe that has a single line of

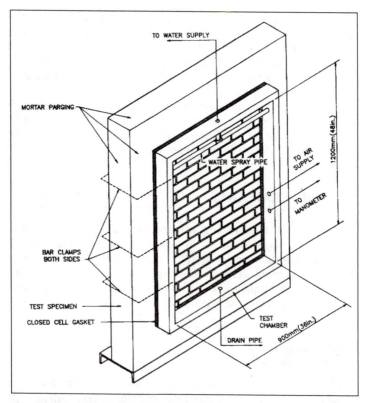

FIGURE 4.4 ASTM E 514 Testing Chamber

holes spaced 1 inch apart. The spray is positioned no more than 3 inches below the top of the chamber. Water is drained from the bottom of the chamber.

Masonry test walls are constructed using materials that are representative of the construction or those being considered for the construction. The height and length of the specimen must be able to accommodate the test chamber while providing a minimum of 8 inches of extension all around the test chamber. The test specimen is built on a steel channel (oriented with the flanges downward) so that the face of the wall that will be exposed to the test is flush with the outside face of one flange of the channel. Bottom-of-specimen flashing, which extends beyond the back of the panel and forms a trough to collect water that passes through the specimen, is provided. Completed specimens are wrapped in plastic for seven days, then cured in laboratory air. A minimum of three test specimens are required per test set.

The chamber is mounted to the test specimen and clamped into place, compressing the gasket seal. Prior to testing, parging is applied to the test face that is outside of the test chamber and to the side and top edges of the specimen. During testing, water is applied at a rate of 3.4 gallons per square foot of wall per hour. Simultaneously, the pressure in the chamber is raised to and maintained at 10 pounds per square foot. The minimum length of the test is 4 hours.

During the test period, observations must be made at 30-minute intervals. Records are kept of:

- Time of appearance of dampness on back of specimen
- Time of appearance of first visible water on the back of the specimen
- Area of dampness on back of wall specimen at the end of the 4-hour test period expressed as a percent of the area tested
- Total water collected from the trough during the 4-hour test period

ASTM E 514 does not provide criteria for evaluating the results of this testing, since the purpose is comparative rather than "pass-fail."

Many investigators have modified the ASTM E 514 test method to accommodate testing in the field rather than in the laboratory. The test chamber, water supply, and pressure conditions are generally the same. However, because the backside of the wall being tested is usually not exposed to view, water on the backside cannot be collected. Instead, a closed system of water supply is used

TABLE 4.6 Suggested Interpretations of Permeance Ratings Determined by Field-Modified ASTM E 514 Testing

Permeance Rating (gal/hr/12 sq. ft.)	Suggested Interpretation
Less than 0.5	Achievable when industry recommendations for workmanship are strictly followed and compatible materials are used.
0.5 to 1.0	Achievable in standard production masonry when industry recognized workmanship recommendations are generally followed and compatible materials are used. Expected range for ordinary brick masonry construction.
1.0 to 2.0	Construction should be considered suspect. In situations where permeance is critical to the serviceability, more detailed investigation is warranted.
2.0 or greater	Construction is considered poor. This rating usually results from workmanship that ignores industry recommendations, wall materials that are not compatible, or both.

(Hoigard, Kudder, & Lies, *Including ASTM E 514 Tests in Field Evaluations of Brick Masonry,* <u>Masonry: Design and Construction, Problems, and Repair,</u> ASTM STP 1180, 1993).

TABLE 4.7 Quality Assurance Requirements: Minimum Tests and Submittals

Level	Minimum Tests and Submittals
1	• Certificates of compliance for materials used in masonry construction
2	• Certificates of compliance for materials used in masonry construction
	• Verification of f'_m prior to construction, except where specifically exempted by Code
3	• Certificates of compliance for materials used in masonry construction
	• Verification of f'_m prior to construction and for every 5000 sq. ft. during construction
	• Verification of proportions of materials in mortar and grout as delivered to the site

(IBC 2000, MSJC Code and Specification 1999, and MSJC Code and Specification 2002).

and monitored. Water that is "lost" from the system is assumed to have penetrated the wall. Measurements of the water supply are made at 30-minute increments. The test is conducted until the rate of water "loss" becomes uniform. Rates of water "loss" that may be considered acceptable are determined by the individual investigator, based on his or her previous experience with using this modified method. Suggested criteria rankings for rates of water loss through walls with a flashing and weep system are presented in Table 4.6.

TABLE 4.8 Quality Assurance Requirements: Minimum Inspection Per MSJC

Level	Minimum Inspection Requirements
1	Verify compliance with the approved submittals
2	As masonry construction begins, verify the following are in compliance:
	• proportions of site-prepared mortar
	• construction of mortar joints
	• location of reinforcement, connectors, and prestressing tendons and anchorages
	• prestressing technique
	Prior to grouting, verify the following are in compliance:
	• grout space
	• grade and size of reinforcement, prestressing tendons, and anchorages
	• placement of reinforcement, connectors, and prestressing tendons and anchorages
	• proportions of site-prepared grout and prestressing grout for bonded tendons
	• construction of mortar joints
	Verify that placement of grout and prestressing grout for bonded tendons is in compliance
	Observe preparation of grout specimens, mortar specimens, and/or prisms
	Verify compliance with the required inspection provisions of the contract documents and the approved submittals

(continued)

TABLE 4.8   Quality Assurance Requirements: Minimum Inspection Per MSJC, *(continued)*

Level	Minimum Inspection Requirements
3	From the beginning of masonry construction and continuously during construction of masonry, verify the following are in compliance:

- proportions of site-prepared mortar, grout, and prestressing grout for bonded tendons
- grade and size of reinforcement, prestressing tendons, and anchorages
- placement of masonry units and construction of mortar joints
- placement of reinforcement, connectors, and prestressing tendons and anchorages
- grout space prior to grouting
- placement of grout and prestressing grout for bonded tendons
- construction of mortar joints

Observe preparation of grout specimens, mortar specimens, and/or prisms

Verify compliance with the required inspection provision of the contract documents and the approved submittals

(MSJC Code and Specification 1999 and MSJC Code and Specification 2002).

QUALITY ASSURANCE

The MSJC and IBC provisions for masonry design and construction require that the quality of masonry materials and workmanship used in the construction be verified. The level of effort that is required to verify the quality of the construction is dependent upon two design parameters: the occupancy use of the building and the structural philosophy used to design the masonry.

Depending upon the occupancy use, the building is designated as either "essential facility" or "non-essential facility" by ASCE 7 "Minimum Design Loads for Buildings and Other Structures." Examples of essential facilities are police stations and hospitals.

Structural philosophies are divided into engineered approaches and prescriptive approaches. Engineered approaches include allowable stress design and strength design. Prescriptive approaches are used in empirical design of masonry and in the design of veneers and glass unit masonry.

TABLE 4.9 Quality Assurance Requirements: Minimum Inspection Per IBC for Level 1

Inspection Task	Freq.*
As masonry construction begins, the following shall be verified to ensure compliance:	
• Proportions of site prepared mortar	P
• Construction of mortar joints	P
• Location of reinforcement and connectors	P
The inspection program shall verify:	
• Size and location of structural elements	P
• Type, size and location of anchors, including details of anchorage of masonry to structural members, frames or other construction	P
• Specified size, grade and type of reinforcement	P
• Welding of reinforcing bars	C
• Protection of masonry during cold weather (below 40°F) or hot weather (above 90°F)	P
Prior to grouting, the following shall be verified to ensure compliance:	
• Grout space is clean	P
• Placement of reinforcement and connectors	P
• Proportions of site-prepared grout	P
• Construction of mortar joints	P
Grout placement shall be verified to ensure compliance	C
Preparation of any required grout specimens, mortar specimens and/or prisms shall be observed	C
Compliance with required inspection provisions of construction documents and approved submittals shall be verified	P

* Frequency of inspection: P = periodically during task listed
C = continuous during task listed
(International Building Code 2000)

MSJC and IBC agree that the lowest level of quality assurance effort applies to prescriptively designed masonry in non-essential facilities. A moderate level of effort is required for prescriptively designed masonry in essential facilities and for engineered masonry in non-essential facilities. The most extensive effort is required only for engineered masonry in essential facilities.

Quality assurance efforts required by IBC and MSJC are defined by two types of requirements: necessary tests and submittals and necessary inspection (called special inspection by IBC). Submittals

TABLE 4.10 Quality Assurance Requirements: Minimum Inspection Per IBC for Level 2

Inspection Task	Freq.*
From the beginning of masonry construction, the following shall be verified to ensure compliance:	
• Proportions of site prepared mortar and grout	P
• Placement of masonry units and construction of mortar joints	P
• Placement of reinforcement and connectors	P
• Grout space prior to grouting	C
• Placement of grout	C
The inspection program shall verify:	
• Size and location of structural elements	P
• Type, size and location of anchors, including details of anchorage of masonry to structural members, frames or other construction	C
• Specified size, grade and type of reinforcement	P
• Welding of reinforcing bars	C
• Protection of masonry during cold weather (below 40°F) or hot weather (above 90°F)	P
Preparation of any required grout specimens, mortar specimens and/or prisms shall be observed	C
Compliance with required inspection provisions of construction documents and approved submittals shall be verified	P

* Frequency of inspection: P = periodically during task listed
C = continuous during task listed
(International Building Code 2000)

and testing that are required by both MSJC and IBC for each level of quality assurance are shown in Table 4.7. Quality assurance Level 1 is the lowest level, and quality assurance Level 3 is the highest.

Although IBC and MSJC agree in spirit about the extent of inspection required in each level of Quality Assurance, details of the inspection requirements differ slightly in those two documents. Table 4.8 lists the inspection requirements as shown in MSJC. Level 2 can be characterized as periodic inspection, while Level 3 is characterized as continuous inspection. In this context, "continuous" is defined to mean that whenever a mason contractor is working on site, an inspector is also required to be on site. It does not mean that every individual masonry unit and every accessory item must be observed as it is placed. In IBC, no special inspection is required for prescriptively designed masonry in non-essential facilities, and IBC (somewhat confusingly) labels the periodic level of inspection as Level 1 and the continuous level of inspection as Level 2. Tables 4.9 and 4.10 list the IBC inspection requirements.

Only the designer-of-record knows the structural philosophy used to design the masonry in the project. Since the appropriate level of inspection is dependent upon the method of structurally designing the masonry, the project documents must identify the required level of quality assurance so that the contractor will know what is expected for the project.

5

INSTALLATION OF MASONRY UNITS

This chapter discusses issues related to installation of masonry units, including planning, unit preparation, bonding, and movement joints. Each of these issues affect the structural integrity of the masonry assembly, as well as the appearance.

PLANNING

Advance planning of masonry unit placement is required so that units will be properly spaced along a length and height of wall. Unit layout should be determined in such a way that corners can be properly bonded and a minimal number of units have to be cut to length at door and window openings.

An initial planning step is to estimate the quantities of units that will be required to complete the construction. The specified sizes of

fastfacts

Unit layout should be determined in such a way that corners can be properly bonded and a minimal number of units have to be cut to length at door and window openings.

fastfacts

Terminology: A single horizontal row of masonry units is called a course. A vertical layer of masonry units that is one unit thick is called a wythe. In construction that utilizes multiple wythes of masonry, the vertical planar space between wythes is called the collar joint. The collar joint may be filled with mortar or grout or be crossed by masonry headers. When the collar joint does not contain mortar, grout, or masonry headers, it can also be called a cavity, even though it may contain rigid insulation.

the masonry units to be used will determine the number of courses needed to make up the required height and the number of units needed to complete the required length of the masonry assembly.

Nominal dimensions of masonry units are normally stated in whole numbers. For modular units, the specified dimensions are less than the nominal dimensions by the thickness of the mortar joints with which the units will be laid. For example, nominal dimensions of a concrete block unit are 8 inches by 8 inches by 16 inches, while its specified dimensions are 7⅝ inches by 7⅝ inches by 15⅝ inches because it is intended to be placed with ⅜-inch-thick mortar joints. Therefore, the modular height of block units is 8 inches and the modular length is 16 inches. Dimensions of modular clay brick units are typically specified so that three courses (three units and three mortar joints) comprise the 8-inch height module and two brick units (two units and two mortar joints) constitute the 16-inch length module. These concepts are illustrated in Figure 5.1.

To assist in planning masonry assemblies consisting of standard modular units, Table 5.1 presents vertical and horizontal coursing dimensions for multiples of brick and block courses. In this table, a

fastfacts

Dimensions of modular clay brick units are typically specified so that three courses (three units and three mortar joints) comprise the 8-inch height module and two brick units (two units and two mortar joints) constitute the 16-inch length module.

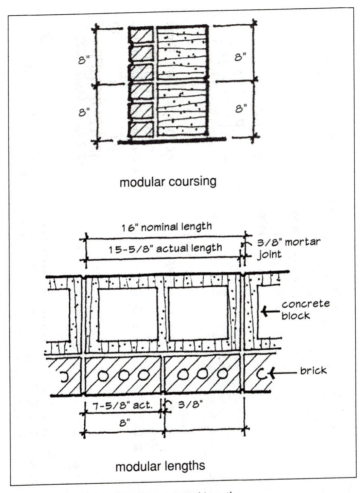

FIGURE 5.1 Modular Unit Coursing and Lengths
(Beall & Jaffe, *Concrete and Masonry Databook*, McGraw-Hill, 2003).

horizontal course is the length of one unit. As demonstrated in Figure 5.1, the length or horizontal course of one block units equals the length or horizontal course of two brick units.

Clay brick units are manufactured in a variety of sizes. Some of these sizes are considered modular, while others are considered non-modular. The height dimension works out to a module of some number of courses to a nominal height for both modular and non-

TABLE 5.1   Modular Unit Coursing

Vertical Coursing			Horizontal Coursing		
No. of brick courses	No. of CMU courses	Wall Height	No. of brick courses	No. of CMU courses	Wall Length
1		2-$\frac{11}{16}$"	1	0.5	8"
2		5-$\frac{5}{16}$"	2	1	1'-4"
3	1	8"	3	1.5	2'-0"
6	2	1'-4"	4	2	2'-8"
9	3	2'-0"	5	2.5	3'-4"
12	4	2'-8"	6	3	4'-0"
15	5	3'-4"	7	3.5	4'-8"
18	6	4'-0"	8	4	5'-4"
21	7	4'-8"	9	4.5	6'-0"
24	8	5'-4"	10	5	6'-8"
27	9	6'-0"	11	5.5	7'-4"
30	10	6'-8"	12	6	8'-0"
33	11	7'-4"	13	6.5	8'-8"
36	12	8'-0"	14	7	9'-4"
39	13	8'-8"	15	7.5	10'-0"
42	14	9'-4"	16	8	10'-8"
45	15	10'-0"	17	8.5	11'-4"
48	16	10'-8"	18	9	12'-0"
51	17	11'-4"	19	9.5	12'-8"
54	18	12'-0"	20	10	13'-4"
57	19	12'-8"	21	10.5	14'-0"
60	20	13'-4"	22	11	14'-8"
63	21	14'-0"	23	11.5	15'-4"
66	22	14'-8"	24	12	16'-0"
69	23	15'-4"	25	12.5	16'-8"
72	24	16'-0"	26	13	17'-4"

(Beall & Jaffe, *Concrete and Masonry Databook,* McGraw-Hill, 2003).

modular brick units, but the length of non-modular units typically does not fit into the standard module. Nominal and specified dimensions, as well as vertical coursing of modular brick units, are presented in Table 5.2, while Table 5.3 presents the same data for non-modular units.

Based on the known face dimensions and area of the masonry units that will be placed in the construction, the number of units that are needed can be estimated from the total square footage of wall

fastfacts

*The MSJC specification requires the contractor to notify the Archi-
tect/Engineer if the bearing of a masonry wythe on its support is
less than two-thirds of hte wythe thickness. For a nominal 4-inch
thick masonry wythe, this means that the units can overhang up
to approximately 1 ¼ inches.*

that will be required. For single-wythe wall construction, the number
of units required can be estimated from Table 5.4. For multi-wythe
wall construction, Table 5.5 can be used to estimate the number of
masonry units that will be required. Table 5.5 is based on composite
wall types illustrated in Figure 5.2, but can also be applied to non-
composite wall types that utilize the same size masonry units.

Planning considerations include inspection of surfaces that will
support masonry construction. Concrete foundation surfaces should
be inspected prior to the start of masonry work to verify that the top
surface is at the correct elevation and is cleaned of laitance, loose
aggregate, dirt, water, and other materials that would inhibit bond
of the mortar to the surface. Even when flashing will be placed on
the supporting surface prior to unit installation, the surface should
be smooth and clean to avoid puncturing the flashing material. Sim-
ilarly, the surface of masonry construction that was completed the
previous day should be inspected to verify that ice or water is not
present. Shelf angles that are attached to the building structure to
periodically support the weight of the outside wythe of masonry
should also be inspected for position. Critical parameters of shelf

fastfacts

*Mortar to masonry unit bond is developed when the unit is able to
suck water out of the mortar and into the pores of the masonry
unit. Cement fines are absorbed into the unit pores along with the
water. Low-suction units may not able to absorb enough water to
adequately develop good bond. High-suction units may absorb too
much water out of the mortar and not leave enough behind to
adequate hydrate the cement and permit it to strengthen.*

TABLE 5.2 Modular Brick Sizes and Vertical Coursing

Unit Designation	Nominal Dimensions (in.)			Joint Thickness (in.)	Specified Dimensions (in.)			Vertical Coursing
	W	H	L		W	H	L	
Modular	4	2⅔	8	⅜ ½	3⅝ 3½	2¼ 2¼	7⅝ 7½	3 courses = 8 in.
Engineer modular	4	3⅕	8	⅜ ½	3⅝ 3½	2¾ 2¹³⁄₁₆	7⅝ 7½	5 courses = 16 in.
Closure modular	4	4	8	⅜ ½	3⅝ 3½	3⅝ 3½	11⅝ 11½	1 course = 4 in.
Roman	4	2	12	⅜ ½	3⅝ 3½	1⅝ 1½	11⅝ 11½	2 courses = 4 in.
Norman	4	2⅔	12	⅜ ½	3⅝ 3½	2¼ 2¼	11⅝ 11½	3 courses = 8 in.
Engineer Norman	4	3¼	12	⅜ ½	3⅝ 3½	2¾ 2¹³⁄₁₆	11⅝ 11½	5 courses = 16 in.

Utility	4	4	12	3/8	3 5/8	3 5/8	11 5/8	1 course = 4 in.
				1/2	3 1/2	3 1/2	11 1/2	
	4	6	8	3/8	3 5/8	5 5/8	7 5/8	2 courses = 12 in.
				1/2	3 1/2	5 1/2	7 1/2	
	4	8	8	3/8	3 5/8	7 5/8	7 5/8	1 course = 8 in.
				1/2	3 1/2	7 1/2	7 1/2	
	6	3 1/5	12	3/8	5 5/8	2 3/4	11 5/8	5 courses = 16 in.
				1/2	5 1/2	2 13/16	11 1/2	
	6	4	12	3/8	5 5/8	3 5/8	11 5/8	1 course = 4 in.
				1/2	5 1/2	3 1/2	11 1/2	
	8	4	12	3/8	7 5/8	3 5/8	11 5/8	1 course = 4 in.
				1/2	7 1/2	3 1/2	1 1/2	
	8	4	16	3/8	7 5/8	3 5/8	15 5/8	1 course = 4 in.
				1/2	7 1/2	3 1/2	15 1/2	

(Beall & Jaffe, *Concrete and Masonry Databook*, McGraw-Hill, 2003). *Continued from page 108.*

TABLE 5.3 Non-Modular Brick Sizes and Vertical Coursing

Unit Designation	Joint Thickness (in.)	W	H	L	Vertical Coursing
Standard	⅜	3⅝	2¼	8	3 courses = 8 in.
	½	3½	2¼	8	
Engineer standard	⅜	3⅝	2¾	8	5 courses = 16 in.
	½	3½	2¹³⁄₁₆	8	
Closure standard	⅜	3⅝	3⅝	8	1 course = 4 in.
	½	3½	3½	8	
King	⅜	3	2¾	9⅝	5 courses = 16 in.
		3	2⅝	9⅝	
Queen	⅜	3	2¾	8	5 courses = 16 in.
	⅜	3	2¾	8⅝	5 courses = 16 in.
		3	2⅝	8⅝	

(Beall & Jaffe, *Concrete and Masonry Databook*, McGraw-Hill, 2003).

TABLE 5.4 Masonry Unit Quantities for Single-Wythe Construction

Nominal Dimensions (in.)		Number of units in
Height	Length	100 s.f. wall area
2	12	600
2⅔	8	676
2⅔	12	450
4	8	450
4	12	300
4	16	225
6	8	300
8	8	225
8	16	112.5

TABLE 5.5 Masonry Unit Quantities for Multi-Wythe Construction

Wall Thickness (in.)	Wall Type per Fig. 5.2	Number of Masonry Units per 100 s.f. of Wall	
		Block	Brick
8	A	112.5 – 4 x 8 x 16	675
	B	97 – 4 x 8 x 16	770
12	C	112.5 – 8 x 8 x 16	675
	D	97 – 8 x 8 x 16	868

(Panarese, Kosmatka, & Randall, *Concrete Masonry Handbook*, PCA, 1991).

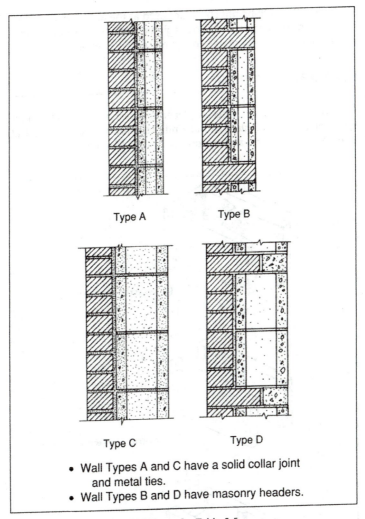

Type A   Type B

Type C   Type D

- Wall Types A and C have a solid collar joint and metal ties.
- Wall Types B and D have masonry headers.

FIGURE 5.2 Composite Wall Types for Table 5.5
(Panarese, Kosmatka, & Randall, *Concrete Masonry Handbook,* PCA, 1991).

angles are not only the elevation but also the location relative to the outside face of masonry and the levelness of the outstanding leg. If the masonry does not adequately overlap the shelf angle to provide enough bearing, the masonry will tend to tilt outward during construction. The outstanding leg of the shelf angle can become out-of-level if less than full-height shims are used at the angle connections. Masonry construction should not begin until the supporting surfaces are deemed to be acceptable to the mason contractor.

UNIT PREPARATION

Preparing the masonry units for placement includes having the right types of units available, cutting units to length where needed at openings or corners, and making sure that the unit has the right

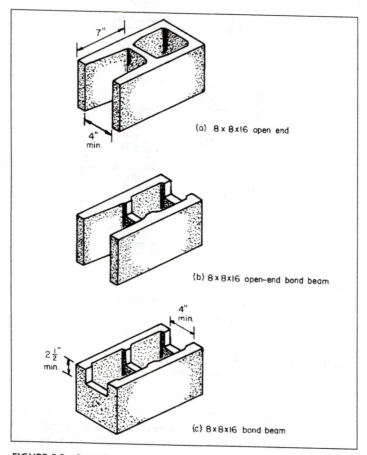

(a) 8 x 8 x 16 open end

7"

4" min.

(b) 8 x 8 x 16 open-end bond beam

4" min.

$2\frac{1}{2}$" min.

(c) 8 x 8 x 16 bond beam

FIGURE 5.3 Specialty Units That Accommodate Reinforcing (Panarese, Kosmatka, & Randall, Concrete Masonry Handbook, PCA, 1991).

amount of moisture to optimize the bond with the mortar. The types of units to be used are specified in the project documents. Specialty forms of those units are often available to make construction of reinforced masonry easier. For example, bond beam units, shown in Figure 5.3 (b) and (c), simplify placement of horizontal reinforcing bars in bond beams and provide sufficient clearance for the grout to surround those bars. Open-end units, seen in Figure 5.3 (a) and (b), make unit placement in vertically reinforced walls easier because the units do not have to be threaded over the top of previously placed reinforcing bars.

Masonry units often have to be cut to size at terminations of the masonry construction. If the designer properly considered the masonry module, cuts are primarily to half the length of the unit and the number of needed cuts is minimized. Units may also have to be halved in width (called a soap) to fit over a protruding anchor bolt that connects a shelf angle, for example.

Clay brick units can be cut with a mason's chisel, called a brick set. While soft brick can often be cut with one sharp blow, harder units must first be scored all around and only then can be severed with a final sharp blow. Concrete and stone units are usually cut with a power saw with a masonry blade. These units may be wetted while sawing to cool the blade and reduce dust. For concrete units, wetting while sawing will also reduce the risk of silicosis. However, these units should be permitted to dry prior to placement in the wall. When concrete units with a high moisture content are placed in the wall, the magnitude of shrinkage and the potential for in-place cracking are increased.

Clay units that have a high initial rate of absorption may have to be prewetted before laying in the wall. Initial rate of absorption, or IRA, is determined by laboratory testing, as described in Chapter 1. Often this value will be given on the test report that is required to be submitted for the project. Mortar generally bonds best to masonry units that have a moderate IRA–from 5 to 25 grams per minute per 30 square inches (the bedding surface area of a standard

fastfacts

Concrete and stone units are usually cut with a power saw with a masonry blade. These units may be wetted while sawing to cool the blade and reduce dust.

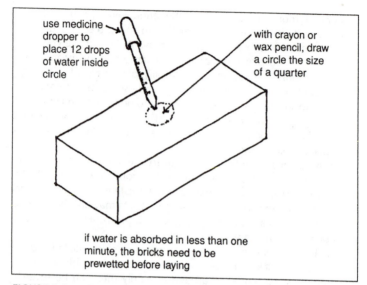

use medicine dropper to place 12 drops of water inside circle

with crayon or wax pencil, draw a circle the size of a quarter

if water is absorbed in less than one minute, the bricks need to be prewetted before laying

FIGURE 5.4   Field Test For Initial Rate of Absorption
(Beall & Jaffe, *Concrete and Masonry Databook,* McGraw-Hill, 2003).

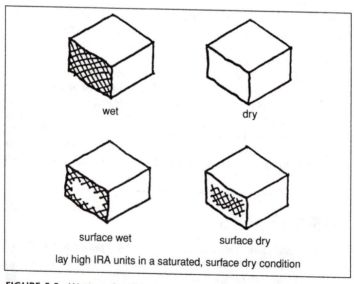

wet

dry

surface wet

surface dry

lay high IRA units in a saturated, surface dry condition

FIGURE 5.5   Wetness Conditions of Masonry Units
(Beall & Jaffe, *Concrete and Masonry Databook,* McGraw-Hill, 2003).

fastfacts

Mortar to masonry unit bond is developed when the unit is able to suck water out of the mortar and into the pores of the masonry unit. Cement fines are absorbed into the unit pores along with the water. Low-suction units may not able to absorb enough water to adequately develop good bond. High-suction units may absorb too much water out of the mortar and not leave enough behind to adequately hydrate the cement and permit it to strengthen.

modular clay brick). Very low suction units tend to float when placed and bond is difficult to achieve. Compatible mortars (those with low water retentivity) should be used with these units. Very high suction units draw too much water from the mortar and create problems with bonding of units placed subsequently. If the IRA is not provided in a test report, a field test procedure, illustrated in Figure 5.4, is available to estimate this value. Using a masonry unit that will be used in the construction, draw a circle the size of a quarter on the bedding surface with a crayon or wax pencil. Using an eye dropper, place 12 drops of water within the circle. If all of the water is absorbed in less than one minute, the brick IRA is too high to lay without prewetting. IBC requires that units with an IRA that exceeds 30 grams per minute per 30 square inches be prewetted.

When clay units have a high IRA, they should be prewetted with a hose until saturated. They should then be permitted to dry until the surface is dry but the interior still is wet. These conditions are shown in Figure 5.5. When saturated but surface dry, high IRA clay units are ready to be placed.

MASONRY UNIT BONDING

Placement of masonry units is critical to the appearance as well as to the structural integrity of the masonry assembly. The way in which the units must overlap with each other within a single wythe or with units in another wythe is dictated by the project documents.

Unit bonding within a wythe is referred to as the bond pattern. Building codes distinguish between two categories of bond patterns: running bond and other than running bond. In running bond, units in each course overlap the units in the courses above and below by at

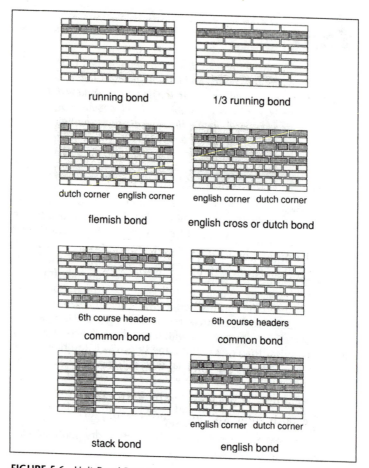

running bond

1/3 running bond

dutch corner   english corner

english corner   dutch corner

flemish bond

english cross or dutch bond

6th course headers

common bond

6th course headers

common bond

stack bond

english corner   dutch corner

english bond

FIGURE 5.6 Unit Bond Patterns
(BIA, Technical Notes on Brick Construction 30, Reissued August 1986).

least one-fourth of the unit length. Typical patterns are one-half running bond and one-third running bond. Other running bond patterns are shown in Figure 5.6. When the units in each course overlap less than one-fourth the unit length, the bond pattern is "other than running bond." Stack bond, in which the head joints between units align vertically, is the most common form of other than running bond.

When the bond pattern is other than running bond, the masonry wythe is required to contain horizontal reinforcement. The minimum area of horizontal reinforcement required by the MSJC Code is

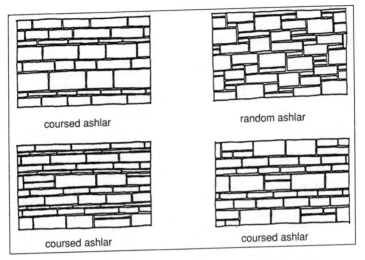

FIGURE 5.7 Ashlar Bonding Patterns
(Beall & Jaffe, *Concrete and Masonry Databook*, McGraw-Hill, 2003).

0.00028 times the vertical gross cross-sectional area of the wall, using specified dimensions. The horizontal reinforcement may consist of joint reinforcement placed in bed joints or of reinforcing bars placed in bond beams spaced at no more than 48 inches on center. When

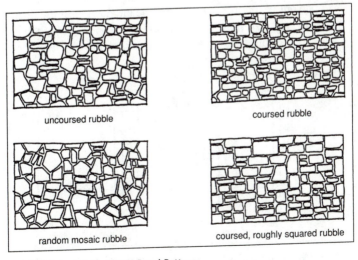

FIGURE 5.8 Rubble Stone Bond Patterns
(Beall & Jaffe, *Concrete and Masonry Databook*, McGraw-Hill, 2003).

the "other than running bond" wythe is designed in accordance with the anchored veneer provisions of the MSJC Code, the veneer is required to have joint reinforcement consisting of at least one wire of size W 1.7 spaced at a maximum of 18 inches on center vertically.

Ashlar bond patterns and rubble bond patterns, commonly used with stone masonry, are shown in Figures 5.7 and 5.8. Architectural concrete masonry units have been successfully laid in ashlar bond patterns to simulate natural stone construction.

Some of the bond patterns shown in Figure 5.6 accommodate the use of masonry headers to connect multiple wythes of masonry. The common bond patterns are most frequently used. As discussed in Chapter 4, construction of multi-wythe walls with masonry headers permits the wall to behave as a structurally composite system. When masonry headers are used to bond together multiple wythes of masonry, the header units are required to be uniformly distributed throughout the wall. The total cross-sectional area of masonry headers must comprise at least 4 percent of the wall surface area, and each header unit must be embedded a minimum of 3 inches into each wythe. These requirements are graphically illustrated in Figure 5.9.

Wall corners are usually bonded for structural continuity. Unless the project documents specifically call for a movement joint at the corner to prevent transfer of shear, the bond pattern should be running bond and the corner should be bonded by one of three methods:

1. 50 percent of the masonry units at the interface must interlock

2. intersecting walls must be anchored by steel connectors that are grouted into the wall at a maximum spacing of 4 feet; min-

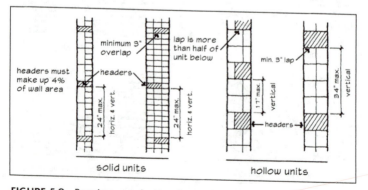

FIGURE 5.9 Requirements for Bonding with Masonry Headers
(Beall & Jaffe, *Concrete and Masonry Databook*, McGraw-Hill, 2003).

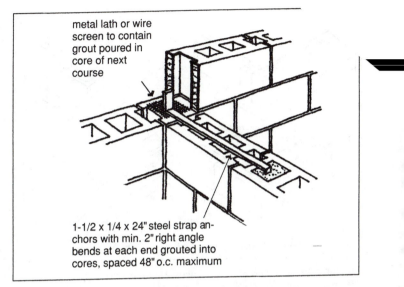

metal lath or wire
screen to contain
grout poured in
core of next
course

1-1/2 x 1/4 x 24" steel strap anchors with min. 2" right angle
bends at each end grouted into
cores, spaced 48" o.c. maximum

FIGURE 5.10 Wall Corner Bonded with Steel Anchor
(Beall & Jaffe, *Concrete and Masonry Databook,* McGraw-Hill, 2003).

imum anchor size ¼ inch by 1½ inches by 28 inches including a
2-inch-long 90-degree bend at each end to form a Z or a U
shape (illustrated in Figure 5.10)

3. intersecting bond beams, reinforced with not less than 0.1
square inch of steel per foot of wall, must be provided at a
maximum spacing of 4 feet; reinforcement must be developed
on each side of the intersection

In order to achieve 50 percent interlock of masonry units at corners, some cutting of units may be required. The cut pieces should
be placed on alternate sides of the wall corner in successive courses
so as to achieve the running bond pattern. Figures 5.11 through 5.14
illustrate how to place units at a wall corner to meet the 50 percent
interlock requirement when the walls are of the same thickness. Figure 5.15 shows methods of providing interlock when the wall thicknesses at a corner are different.

Masonry construction usually begins at a wall corner. If a movement joint is specified to be placed near the corner, the masonry can
be terminated in a vertical line at the movement joint location.
When a movement joint is not specified to be placed near the corner, however, masonry units should be placed so that the temporary

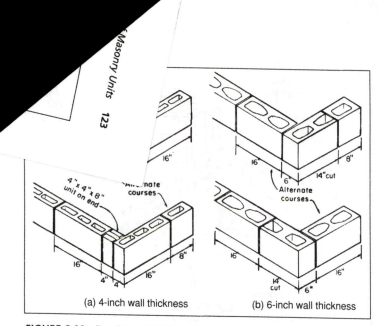

FIGURE 5.11 Bonding of Wall Corners—Uniform Thickness
(Panarese, Kosmatka, & Randall, *Concrete Masonry Handbook*, PCA, 1991).

stopping of construction forms a racked back condition rather than a toothed condition. When masonry construction resumes where the masonry is racked back, as shown in Figure 5.16, the new masonry units are placed on top of existing construction. When units have to be placed at a toothed wall termination, however, the units have to be slid between existing projecting units in the courses above and below. Because it is difficult to achieve good bond at units that are placed into toothed construction, this condition should be avoided.

Considerable visual interest can be given to a masonry wall by recessing and projecting courses of masonry units. Several adjacent

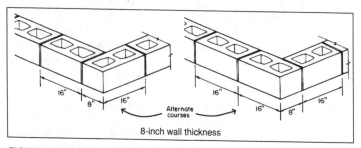

FIGURE 5.12 Bonding of Wall Corners—Uniform Thickness
(Panarese, Kosmatka, & Randall, Concrete Masonry Handbook, PCA, 1991).

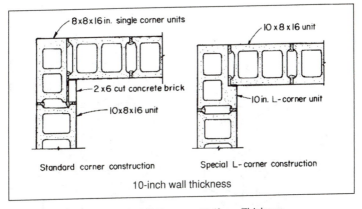

FIGURE 5.13 Bonding of Wall Corners—Uniform Thickness
(Panarese, Kosmatka, & Randall, *Concrete Masonry Handbook,* PCA, 1991).

courses may be successively projected to form a corbel, illustrated in Figure 5.17. There are code limitations on the maximum amount that individual courses may project and on the total projection of multiple courses. The maximum projection of one unit is not permitted to exceed one-half the unit height or one-third the unit thick-

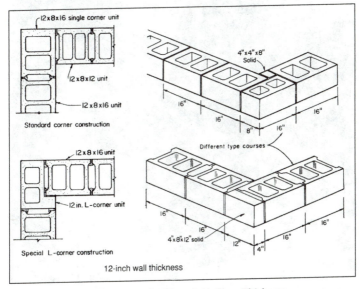

FIGURE 5.14 Bonding of Wall Corners—Uniform Thickness
(Panarese, Kosmatka, & Randall, *Concrete Masonry Handbook,* PCA, 1991).

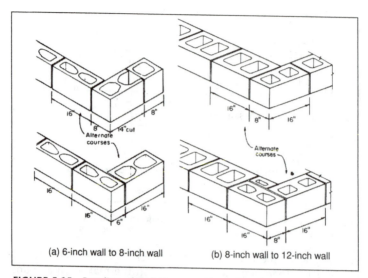

FIGURE 5.15 Bonding of Wall Corners—Nonuniform Thickness
(Panarese, Kosmatka, & Randall, *Concrete Masonry Handbook,* PCA, 1991).

Labels in figure: (a) 6-inch wall to 8-inch wall; (b) 8-inch wall to 12-inch wall; Alternate courses; 8" 14" cut; 16"; 8"; 6"

FIGURE 5.16 Racked Back Masonry at a Corner
(photo courtesy of Portland Cement Association).

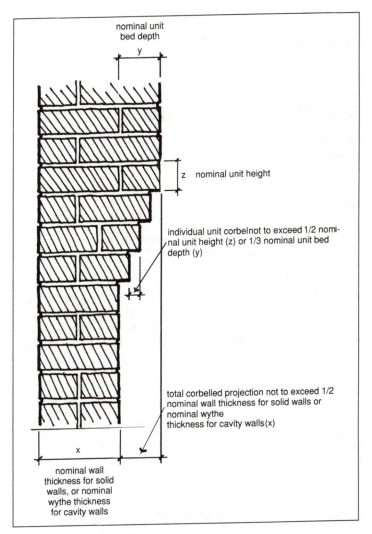

nominal unit
bed depth

y

z nominal unit height

individual unit corbelnot to exceed 1/2 nominal unit height (z) or 1/3 nominal unit bed
depth (y)

total corbelled projection not to exceed 1/2
nominal wall thickness for solid walls or
nominal wythe
thickness for cavity walls(x)

x

nominal wall
thickness for solid
walls, or nominal
wythe thickness
for cavity walls

FIGURE 5.17 Corbelled Masonry Requirements
(Beall & Jaffe, *Concrete and Masonry Databook*, McGraw-Hill, 2003).

ness measured at right angles to the wall. The total corbelled projection is not permitted to exceed one-half the wall thickness for
solid walls or one-half the wythe thickness for hollow walls. Solid
masonry units must be used when corbelling. When constructing a
corbel, additional pieces of masonry units must be installed behind

Movement joints may be generally categorized into three types: expansion joints, control joints, and isolation joints.

the projecting units so that the opposite side of the wall or wythe remains in the same plane.

MOVEMENT JOINTS

Movement joints are necessary to control the potentially adverse effects of volume changes in masonry construction. The movement joints permit the dimensional changes to take place in locations that are planned so that neither the structural nor the aesthetic performance of the masonry is reduced. The MSJC Code requires that the designer include "provision for dimensional changes resulting from elastic deformation, creep, shrinkage, temperature, and moisture" in the project drawings. In the field, movement joints are constructed where shown by the designer-of-record.

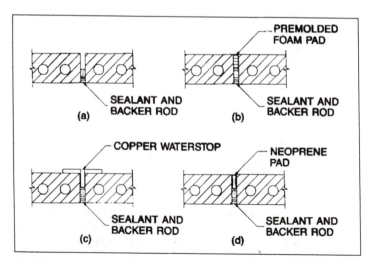

FIGURE 5.18   Expansion Joints in Clay Masonry
(BIA Technical Note 5 on Brick Construction 18A, December 1991).

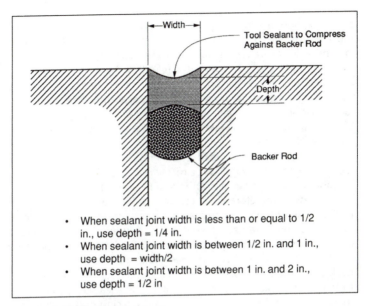

- When sealant joint width is less than or equal to 1/2 in., use depth = 1/4 in.
- When sealant joint width is between 1/2 in. and 1 in., use depth = width/2
- When sealant joint width is between 1 in. and 2 in., use depth = 1/2 in

FIGURE 5.19 Sealant Joint Profile

Movement joints may be generally categorized into three types: expansion joints, control joints, and isolation joints. Expansion joints are used in clay masonry construction to provide an open space into which clay units can grow. Clay units experience irreversible moisture expansion over their lifetime, most of which occurs during the first year after being removed from the kiln. All moisture is removed from these units during the manufacturing process, and the units slowly absorb moisture from the air once they have left the kiln.

Typical expansion joint details are shown in Figure 5.18. Details 5.18(a) and 5.18(b) are the most commonly used forms of expansion joints. Details 5.18(c) and 5.18(d) are used when more protection against water penetration is desired. The profile of the sealant used in these joints should always have an hourglass shape, with the depth or thickness of the sealant equal to about half the sealant joint width, as shown in Figure 5.19. Thicker sealant joints are too stiff to properly accommodate movement and thinner sealant joints have inadequate strength to resist movement. The hourglass shape is accomplished by installing the sealant against a foam backer rod and tooling the sealant surface with a concave tool.

Clay brick masonry grows both horizontally and vertically. Therefore, expansion joints must be provided in both directions. At horizontal expansion joints, the weight of the masonry above the joint must be supported by a structural member. Therefore, horizontal expansion joints are typically shown to be placed below shelf angles, as shown in Figure 5.20. Unfortunately, this creates an overall joint that is considerably wider than the typical mortar bed joint, because the expansion must be accommodated below the shelf angle. The designer-of-record is therefore challenged in detailing this joint to be both aesthetically pleasing yet capable of accommodating the anticipated movement.

Control joints are provided in concrete masonry to control the location of the shrinkage crack that will inevitably form. Concrete masonry undergoes net shrinkage as a result of the curing and drying processes. Therefore, it is necessary to form a weakened section in the plane of the concrete masonry wythe so that the crack location can be predicted and forced to occur in an aesthetically pleasing manner. A control joint can be formed by simply omitting mortar from vertically aligned head joints and placing sealant and backer rod into those joints instead. However, it is normally desirable to have a shear key at the control joint. The shear key prevents the masonry on one side of the joint from displacing out-of-plane relative to the masonry on the opposite side of the joint. From a structural perspective, the shear key also permits transfer of out-of-plane

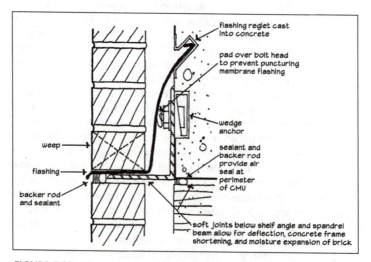

FIGURE 5.20   Horizontal Expansion Joint
(Beall & Jaffe, *Concrete and Masonry Databook*, McGraw-Hill, 2003).

FIGURE 5.21 Control Joint Formed with Neoprene Gasket
(photo courtesy of Portland Cement Association).

forces across the joint so that the masonry need not be considered laterally unsupported at the joint.

There are three common methods of forming control joints with shear keys in concrete masonry. One method is to use special units that are fabricated with a recess in the square end to receive a neoprene gasket. This detail is shown in Figure 5.21. A second method is to use special units that are fabricated to interlock. The end of one special unit has a wide notch and the end of the other special unit has a wide projection that fits into the notch, as seen in Figure 5.22. In both of these control joint details, no mortar is installed at the joint; sealant and backer rod are installed at each face of the joint. The third method utilizes regular concave end units. A bond breaker, such as building paper, is placed against the end of one unit and grout is placed into the space between units. Mortar that was placed along the face shells to contain the grout is raked back and a bond breaker and sealant are installed on each face. This joint detail, illustrated in Figure 5.23, is commonly called a Michigan control joint.

Regardless of which method is used to form the control joint in concrete masonry or the expansion joint in brick masonry, joint reinforcement should not be placed continuously across the move-

FIGURE 5.22 Control Joint Formed with Special Units
(photo courtesy of Portland Cement Association).

FIGURE 5.23 Michigan Control Joint
(photo courtesy of Portland Cement Association).

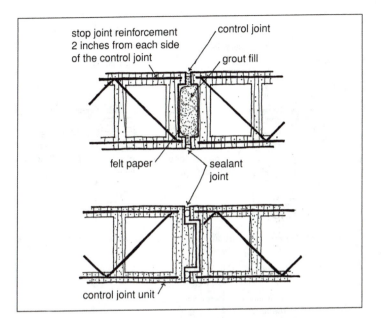

FIGURE 5.24 Discontinue Joint Reinforcement at Control Joint
(Beall & Jaffe, *Concrete and Masonry Databook,* McGraw-Hill, 2003).

ment joint. Joint reinforcement wires should be cut approximately 2 inches from each side of the joint, as shown in Figure 5.24.

Isolation joints are used to provide a separation between masonry and other elements of construction, such as windows, doors, or the structural frame. Isolation joints permit the different materials to move independently of each other and prevent transfer of structural forces into non-structural elements. At locations where lateral support of the masonry is required, but the masonry also needs to be isolated from the vertical deflection and loading supported by the structural element, an isolation joint should be provided. Examples of this type of joint are shown in Figure 5.25.

TOLERANCES

Like all forms of construction, achievement of perfection is not a reasonable expectation. Some degree of difference between the requirements stated in the project documents and actual construction is to be expected. Codes define the amount of difference, called

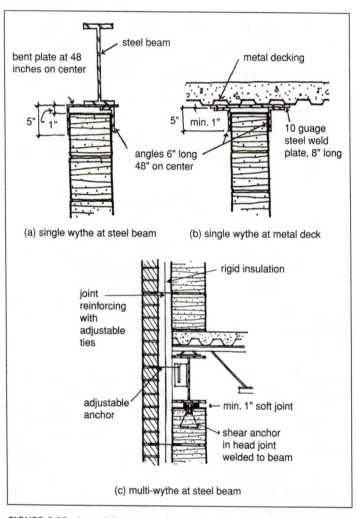

steel beam

bent plate at 48
inches on center

5" 1"

angles 6" long
48" on center

metal decking

5" min. 1"

10 guage
steel weld
plate, 8" long

(a) single wythe at steel beam

(b) single wythe at metal deck

rigid insulation

joint
reinforcing
with
adjustable
ties

adjustable
anchor

min. 1" soft joint

shear anchor
in head joint
welded to beam

(c) multi-wythe at steel beam

FIGURE 5.25 Lateral Support and Isolation at Tops of Non-Bearing Walls
(Beall & Jaffe, *Concrete and Masonry Databook*, McGraw-Hill, 2003).

tolerance, that is acceptable from a structural point of view. The
project documents may require stricter tolerances where uniform
appearance is of utmost importance.

The MSJC Specification defines acceptable tolerances for masonry
elements relative to location, dimension, level, plumb, and true to a

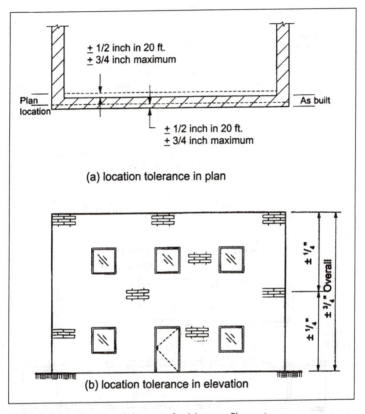

FIGURE 5.26   Location Tolerances for Masonry Elements
(Chrysler & Amrhein, *Reinforced Concrete Masonry Construction Inspector's Handbook*, Masonry Institute of America, 2002).

line. Relative to plan location, masonry elements are required to be built within plus or minus ½ inch in 20 feet, with a maximum difference of plus or minus ¾ inch. Relative to elevation location, masonry elements are permitted to vary plus or minus ¼ inch per story height, with a maximum variation of plus or minus ¾ inch. Tolerances in location of masonry elements are shown in Figure 5.26.

Tolerance requirements for dimensions of masonry elements in cross-section or in elevation, according to the MSJC Specification, is minus ¼ inch or plus ½ inch. The MSJC Specification also gives requirements for permitted variations from level. Bed joints and the top surface of bearing walls are required to be level within plus or minus ¼ inch in 10 feet, with a maximum out-of-level of ½ inch. Figure 5.27 illustrates level tolerances.

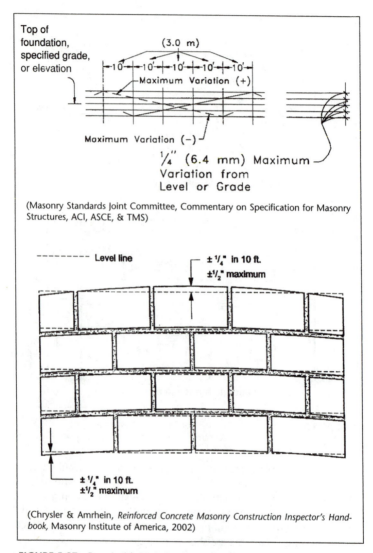

(Masonry Standards Joint Committee, Commentary on Specification for Masonry Structures, ACI, ASCE, & TMS)

(Chrysler & Amrhein, *Reinforced Concrete Masonry Construction Inspector's Handbook*, Masonry Institute of America, 2002)

FIGURE 5.27 Permissible Variation From Level for Bed Joints or Top Surface of Bearing Walls

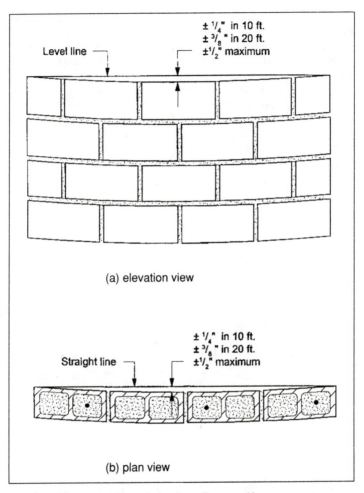

FIGURE 5.28 Permissible Variation From True to a Line
(Chrysler & Amrhein, *Reinforced Concrete Masonry Construction Inspector's Handbook*, Masonry Institute of America, 2002)

According to the MSJC Specification, masonry elements are permitted to vary from plumb (true vertical plane) by the following amounts:

- plus or minus ¼ inch in 10 feet;
- plus or minus ⅜ inch in 20 feet; or
- plus or minus ½ inch maximum.

The same tolerance limitations apply to constructing masonry elements true to a line, which may be a straight line in plan view or a level line in elevation view. These tolerances are shown in Figure 5.28.

Finally, the MSJC Specification states allowable tolerances for alignment of masonry columns and walls. These tolerances apply to the bottom of the element when compared to the top of the element. The permitted tolerances are plus or minus ½ inch for bearing walls and plus or minus ¾ inch for non-bearing walls.

MORTAR AND GROUT INSTALLATION

Mortar is the material that binds together masonry units. It also separates the units, permitting level installation of courses. Mortar further provides the function of sealing the gap between masonry units and preventing rainwater from entering that gap. Proper performance of the masonry assembly, therefore, depends upon proper placement of the mortar when laying the units.

Grout is used to bind steel reinforcement and connectors to the masonry. It can also be used to fill the collar joint between masonry wythes. The addition of grout to a masonry wall increases its structural capacity, fire resistance, sound transmission rating, and ability to prevent water passage through the wall.

This chapter discusses the proper placement of mortar and grout in masonry construction.

fastfacts

Proper performance of the masonry assembly depends upon proper placement of the mortar when laying the units.

MORTAR PLACEMENT IN NEW CONSTRUCTION HEAD JOINTS AND BED JOINTS

When mixing mortar in preparation for placement, proper proportioning of ingredients is critical. While the volume of prepackaged cement, lime, and admixtures is relatively easy to control, sand volume is not. Sand increases in bulk with increases in moisture content so that the volume of a given quantity of sand may vary throughout the day and from day to day. Table 6.1 demonstrates the variation in sand volume with variation in water content.

Measuring sand by counting shovels is not an accurate method of batching and is not recommended. Two alternate methods for accurately measuring sand volume are illustrated in Figure 6.1. In Figure 6.1a, measuring boxes are being used to check the number of shovels of sand it takes to fill one cubic foot. Measuring boxes should be used at least twice a day to check sand volume, once in the morning and again after lunch. Figure 6.1b shows a 1-cubic-foot batching box in which the sand is shoveled and then discharged into the mixer directly. The second method is more accurate and accounts for continuous volume changes in the sand as it dries or bulks with moisture changes.

All of the aggregate and cementitious materials (cement and lime) should be mixed for three to five minutes in a mechanical batch mixer with the *maximum* amount of water to produce a workable consistency. Unlike concrete, water content is not limited in field-mixed mortar. Limitations are not imposed by the standards because water content of mortar is self-regulating: if there is too much water, the mortar will not be stiff enough to support the weight of the units as they are laid; and if there is too little water, the mortar will not be workable and will not be easily spread.

Within the first 2½ hours after mixing, mortar that stiffens because of evaporation should be retempered by adding water.

TABLE 6.1 Volume Measurement of Mortar Materials

Material	Volume (cu. ft.)
1 bag Portland cement	1.0
1 bag hydrated lime	1.0
1 ton wet sand	20.25
1 ton damp, loose sand	18.25
1 ton dry sand	16.25

(Beall & Jaffe, *Concrete and Masonry Databook*, McGraw-Hill, 2003).

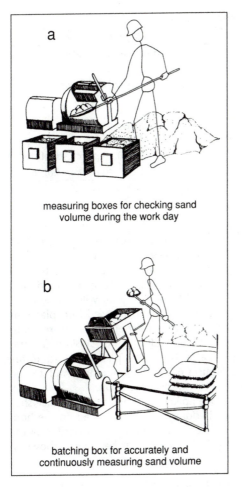

measuring boxes for checking sand
volume during the work day

batching box for accurately and
continuously measuring sand volume

FIGURE 6.1 Accurate
Batching of Sand for
Mortar and Grout
(Beall and Jaffe, *Concrete
and Masonry Databook,*
McGraw-Hill, 2003).

Retempering is permitted as frequently as needed to restore the
required consistency to the mortar. This will assure better mortar
bond by keeping workability at the optimum. However, when colored additives are used, the mortar should not be retempered
because color variations will result. Unused mortar should be dis-
carded beyond 2½ hours after mixing. Tables 6.2 through 6.4 are
provided as guides for estimating the quantity of mortar that will be
needed in a project.

Solid masonry units (those with core holes comprising no
more than 25 percent of the cross-section) are placed in full beds

TABLE 6.2   Estimating Mortar Quantities

Type of Masonry	Mortar Quantity (cu.yds.)
3⅝" x 2¼" x 7⅝" modular brick with ⅜" mortar joints	0.515 per 1000 brick
Nominal 8" x 8" x 16" concrete block with ⅜" mortar joints	1.146 per 1000 block
4" x 1½" x 8" paving brick	
⅜" mortar joints	0.268 per 1000 pavers
1" thick mortar setting bed	0.820 per 1000 pavers
Cut stone	0.04 to 0.10 per cu.yd. of stone
Fieldstone	0.15 to 0.40 per cu.yd.of stone

(Adapted from Kolkoski, *Masonry Estimating*, Craftsman Book Co., 1988).

(horizontal joints) of mortar and the head (vertical) joints are completely filled. This practice is known as full mortar bedding. Sometimes the masonry contractor will furrow the bed joint to ease unit placement. Deep furrowing, or using the trowel to form a channel in the middle of the bed joint, is not recommended. If the furrowing is more than slight, the bed joint will not be completely filled with mortar after the unit is pressed into place, as seen in Figure 6.2. Ends of masonry units should be buttered with mortar prior to placement in the wall and not slushed (mortar thrown into the head joint) after the unit is placed. Less-than-full head joints, as shown in Figures 6.3 and 6.4, and incomplete bed joints are more susceptible to water penetration.

Industry practice is to place hollow masonry units with face shell bedding only. In face shell bedding, mortar is placed on the horizontal and vertical face shells only, leaving the inner portion of the head and bed joints open. Face shell mortar bedding is shown in

TABLE 6.3   Estimated Mortar Quantities per 100 Square Feet of Wall Surface

Wall Thickness (in.)	Wall Type	Mortar (cu. ft.)
4, 6, or 8	Single Wythe–4 in. high units	13.5
4, 6, 8, or 12	Single Wythe–8 in. high units	8.5
8	Composite–A*	20.0
8	Composite–B*	12.2
12	Composite–C*	20.0
12	Composite–D*	13.5

* Wall Types are illustrated in Figure 5.2

(Panarese, Kosmatka, & Randall, *Concrete Masonry Handbook*, PCA, 1991).

TABLE 6.4 Sample Mortar Quantities

Mortar Type	Mortar Mix Proportions, parts by volume				Material Quantities, cu. ft. for 1 cu. ft. of mortar			
	Portland Cement	Masonry Cement	Hydrated Lime	Sand	Portland Cement	Masonry Cement	Hydrated Lime	Sand
M	1	1	–	6	0.16	0.16	–	0.97
M or S	1	–	¼	3	0.29	–	0.07	0.96
N	1	–	1	6	0.16	–	0.16	0.97
N or O	–	1	–	3	–	0.33	–	0.99

(Panarese, Kosmatka, & Randall, *Concrete Masonry Handbook*, PCA, 1991).

FIGURE 6.2 Incompletely Filled Bed Joints
(The Magazine for Masonry Construction, May 1990).

Figure 6.5. Cross webs of hollow masonry units are covered with mortar only at specific locations:

- All courses of piers, columns, and pilasters
- First course on the foundation
- Adjacent to vertical cells scheduled for grouting in partially grouted walls
- When necessary to confine loose fill insulation

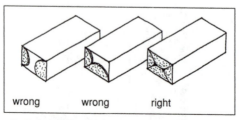

FIGURE 6.3 Buttering of Head Joints
(*Technical Notes on Brick Construction* 7B, Brick Industry Association, April 1998).

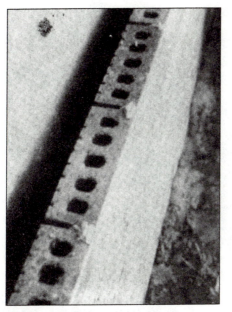

FIGURE 6.4 Incompletely Filled Head Joints
(The Magazine for Masonry Construction, May 1990).

Also, the designer/specifier may require mortaring of all cross webs (full mortar bedding) if additional structural capacity is needed in the wall. Mortared cross webs at a pilaster are shown in Figure 6.6.

Mortar joint width depends upon the type of masonry unit utilized. Modern modular units typically allow for a ⅜-inch-wide mortar joint. Construction variations will occur, and codes specify the permitted tolerance for mortar joint widths. Tolerance limitations imposed by the MSJC Specification are shown in Table 6.5. Head joint widths are permitted to vary more than bed joints, because head joint tolerances will accumulate over the length of the masonry course. A standard ⅜-inch-thick (as specified) bed joint is acceptable as long as it is actually between ¼-inch and ½-inch thick. A standard ⅜-inch-thick (as specified) head joint, however, can be between ⅛-inch and ¾-inch thick and still be acceptable. These tolerances are based on structural capacity and not visual effects. A designer/specifier may require more stringent tolerance limitations for a particular project in order to achieve the desired aesthetic appearance.

FIGURE 6.5 Face Shell Mortar Bedding
(Photograph courtesy of Portland Cement Association).

Mortar joints should be tooled when the mortar has become thumbprint hard (the thumbprint will not be visible on the mortar surface). Tooling densifies the mortar surface and brings the mortar into tight contact with the adjacent masonry units. These qualities make the joint more resistant to water penetration.

FIGURE 6.6 Mortared Cross Webs at Pilaster
(Photograph courtesy of Portland Cement Association).

TABLE 6.5 Mortar Joint Thickness Tolerances

| Joint Type | Tolerance, in. | |
	Plus	Minus
bed	$1/8$	$1/8$
head	$3/8$	$1/4$
collar	$3/8$	$1/4$

(Specification for Masonry Structures ACI 530.1/ASCE 6/TMS 602, Masonry Standards Joint Committee, 2002).

MORTAR PLACEMENT IN EXISTING CONSTRUCTION HEAD JOINTS AND BED JOINTS

When mortar joints in existing construction become weathered or otherwise deteriorated, repair consists of replacing the outer portion of the mortar. The process of removing deteriorated mortar and installing new mortar is called repointing. Tuckpointing is the placement of new mortar over existing mortar, and pointing is the process of installing new mortar into a prepared joint. However, tuckpointing is commonly used to mean repointing.

In masonry construction of solid units, industry recommendations are to remove mortar to a minimum of ⅜-inch deep until solid mortar is reached, up to a maximum depth of half the masonry unit depth. In hollow unit masonry, the depth of mortar removal should be limited to half the face shell thickness. If more mortar is removed, the risk of breaking through the mortar joint to the open cell interior of the unit is increased. If some existing mortar is not left in the back of the mortar joint, there will not be a backing against which the new mortar can be compressed as it is installed and it will be difficult to fill the joint. Mortar should be removed from head and bed joints to form a square profile, exposing the surface of the masonry units at

fastfacts

Tuckpointing is the placement of new mortar over existing mortar, and pointing is the process of installing new mortar into a prepared joint. However, tuckpointing is commonly used to mean repointing: removal of deteriorated mortar and installing new mortar.

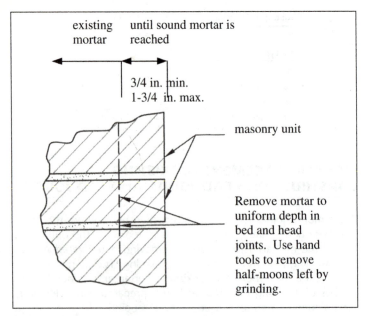

FIGURE 6.7 Mortar Removal for Repointing

each interior surface of the joint but not damaging the masonry units. Refer to Figure 6.7. Exposing the unit surface within the joint gives the repointing mortar the opportunity to bond with the units. After mortar removal is complete, the prepared joints should be cleaned of loose particles, dust, and debris using an air jet or a water stream.

Mortar that is used for repointing existing masonry must be carefully selected. It must be compatible with the original mortar and never harder, denser, or less absorbent. Ideally, original mortar should be examined microscopically and analyzed chemically to determine

fastfacts

The standard procedure for analyzing hardened mortar is provided in ASTM C 1324 "Standard Test Method for Examination and Analysis of Hardened Masonry Mortar." This standard defines how to examine the mortar sample petrographically (microscopically) and chemically to define its constituent materials and proportions.

TABLE 6.6 ASTM-Recommended Repointing Mortars

	Mortar Type	
Location	Recommended	Alternative
Interior	O	K, N
Exterior, above grade exposed on one side, unlikely to be frozen when saturated, not subject to high wind or other significant lateral load	O	N, K
Exterior, other than above	N	O

(Extracted, with permission, from ASTM C 270 *Standard Specification for Mortar for Unit Masonry,* copyright ASTM International, 100 Barr Harbor Drive, West Conshohocken, PA, 19428. A copy of the complete standard may be obtained from ASTM, phone: 610-832-9585m email: service@astm.org, website: www.astm.org).

its makeup and permit formulation of a compatible repointing mortar. This procedure is especially important when repointing historic structures. Guidelines for selection of a repointing mortar mix given by ASTM are provided in Table 6.6, and those given by the Ontario Ministry of Citizenship and Culture are in Table 6.7.

Mortar for repointing is prepared differently than mortar for new construction. Generally, it is mixed to a drier consistency and is permitted to prehydrate prior to placement. Prehydration consists of adding water to the dry mortar component materials and letting the mixture stand for 1 to 1½ hours. This process permits some of the shrinkage, which results when the cement hydrates or undergoes its chemical change, to occur prior to the mortar being placed in the wall. Therefore, mortar shrinkage after placement is reduced. The mixing procedure outlined in ASTM C 270 for preparing repointing mortar is given in Table 6.8.

Repointing mortar should be used within 2½ hours of initial mixing. Therefore, repointing mortar only has a board life of 1 to 1½ hours, compared to 2½ hours for mortar for new construction. The prehydration time is deducted from the usable time period. Like mortar for new construction, uncolored repointing mortar is permitted to be retempered within the allowable time period.

A special tuckpointing trowel, which is narrow enough to fit in the joints between masonry units, is used to place repointing mortar into the ground-out joints. Repointing mortar should be placed in lifts no more than ¼ inch deep. Repointing lifts are illustrated in Figure 6.8. The Brick Industry Association[1] recommends that each lift be permitted to become thumbprint hard before placing the

[1] Technical Note 7F, Moisture Resistance of Brick Masonry – Maintenance, Brick Industry Association, Reissued January 1987 (Reissued October 1998)

TABLE 6.7 Canadian-Recommended Repointing Mortars

Masonry Material	Recommended Mortar Mix Proportions for Various Weathering Exposures		
	Sheltered	Moderate	Severe
Highly durable granite or hard brick	1 part cement 2 parts lime 8 to 9 parts sand	1 part cement 1½ parts lime 5 to 6 parts sand	1 part cement ½ part lime 4 to 4½ parts sand
Moderately durable stone or brick	1 part cement 3 parts lime 10 to 12 parts sand	1 part cement 2 parts lime 8 to 9 parts sand	1 part cement 1½ parts lime 5 to 6 parts sand
Poorly durable soft brick or friable stone	0 parts cement 2 parts hydraulic lime[a] 5 parts sand	1 part cement 3 parts lime 10 to 12 parts sand	1 part cement 2 parts lime 8 to 9 parts sand

[a] A product that hardens and gains strength over time; different than hydrated lime.
(Adapted from the Ontario Ministry of Citizenship and Culture, *Annotated Master Specifications for the Cleaning and Repointing of Historic Masonry Structures,* 1985).

TABLE 6.8 Mixing Sequence for Repointing Mortar

Step	Action
1	Dry mix all dry (sand and cementitious) materials.
2	Add only enough water to produce a damp mix that will retain its shape when pressed into a ball by hand.
3	Mix for at least three (3) and not more than seven (7) minutes in a mechanical mixer.
4	Permit mixed mortar to stand for a minimum on one (1) hour and a maximum of 1½ hours for pre-hydration.
5	Add sufficient water to bring mortar to proper consistency for tuckpointing, somewhat drier than mortar used for laying masonry units.
6	Mix by hand for three (3) to five (5) minutes.

(Extracted, with permission, from ASTM C 270 *Standard Specification for Mortar for Unit Masonry*, copyright ASTM International, 100 Barr Harbor Drive, West Conshohocken, PA, 19428. A copy of the complete standard may be obtained from ASTM, phone: 610-832-9585m email: service@astm.org, website: www.astm.org).

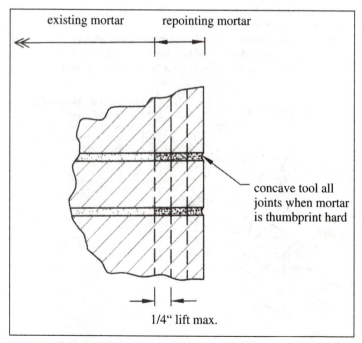

FIGURE 6.8 Installation of Repointing Mortar in Lifts

next lift. Others believe that it is better to place the successive lift without delay, since the mortar will bond better to fresh mortar than to partially hardened mortar in the previous lift.

COLLAR JOINT MORTARING

The collar joint is the three-dimensional space between two wythes (or layers) of masonry. The collar joint may be filled with mortar or grout, making the wall structurally composite, or void of mortar and grout, making the wall structurally non-composite. A structurally non-composite wall is also known as a cavity wall. The collar joint in a non-composite wall may contain insulation.

When the collar joint is specified to be open, mortar extrusions and droppings in the cavity must be avoided. Extrusions that bridge across the cavity to the inside masonry wythe permit water that penetrates the outside wythe to travel across the cavity and enter the interior wythe. From there, the water can access interior finishes that are easily damaged. Mortar droppings on the flashing inhibit water flow out of the wall, increasing the likelihood of efflorescence and other moisture-related problems.

The best way to prevent or at least minimize mortar extrusions and droppings is to bevel the mortar bed joint, as illustrated in Figure 6.9. When the masonry unit is placed on a beveled mortar bed, the mor-

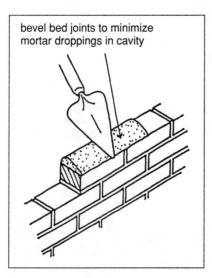

bevel bed joints to minimize mortar droppings in cavity

FIGURE 6.9 Beveling of Mortar Bed Joints
(*Technical Notes on Brick Consruction* 7B, Brick Industry Association, April 1998).

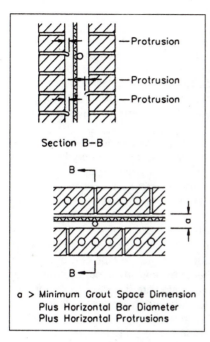

FIGURE 6.10 Clear Width Dimension in Collar Joint (Beall and Jaffe, *Concrete and Masonry Databook,* McGraw-Hill, 2003).

tar extrudes toward the outside of the wall where the mason can easily remove it with the trowel and not toward the cavity.

When the collar joint is specified to be filled, it must be completely filled. A partially filled collar joint is an ineffective barrier to rain penetration. Depending upon the width of the collar joint, it may be filled with mortar or grout.

When the collar joint has less than ¾ inch clear width, either due to its specified dimension or mortar extrusions, mortar must be used to fill the joint. When using mortar to fill a collar joint, mortar is placed in the joint as the units are laid. Collar joint mortar is placed after one wythe is already in place and the second wythe is being placed (either one course at a time or full wall height). Either the back of the facing unit being placed may be buttered, or the face of the back-up unit already in place may be parged. Slushing of the collar joint is not recommended because it results in incomplete filling.

Collar joints that are wider than ¾ inch clear may be filled with grout. The clear width is exclusive of mortar protrusions and diameters of horizontal reinforcing bars placed in the collar joint. Refer to Figure 6.10. Grout placement is discussed in the next section.

GROUT PLACEMENT

Grout may be placed in the collar joint between masonry wythes to form a structurally composite wall (and functional barrier wall) or in the cells of hollow units. Grout may be placed with or without steel reinforcement to enhance the wall's properties, but grout must always be used wherever the masonry contains steel reinforcing bars.

Measures for ensuring consistency in proportions of grout materials, particularly sand, are the same as those undertaken during mortar preparation. Measuring boxes or batching boxes, shown in Figure 6.1, should be used. Sufficient water should be added to produce grout with slump between eight inches and eleven inches, as discussed in Chapter 2.

The maximum grout pour height (height to which the masonry is constructed prior to grout placement) is based on the clear dimension of the space to be grouted and the type of grout (fine or coarse). Grout space requirements and maximum grout pour heights given in the MSJC Specification are shown in Table 6.9. Figure 6.10 demonstrates the clear dimension requirements for collar joints between wythes, and the grout space requirements for cells or cores of hollow units are shown in Figure 6.11.

TABLE 6.9 Grout Space Requirements

Grout Type	Maximum Grout Pour Height ft. (m)	Minimum Width of Grout Space Between Wythes of Masonry[1] in. (mm)	Minimum Grout Space Dimensions for Grouting Cells or Cores of Hollow Units[1,2] in. x in. (mm x mm)
Fine	1 (0.30)	¾ (19.1)	1½ x 2 (38.1 x 50.8)
Fine	5 (1.52)	2 (50.8)	2 x 3 (50.8 x 76.2)
Fine	12 (3.66)	2½ (63.5)	2½ x 3 (63.5 x 76.2)
Fine	24 (7.32)	3 (76.2)	3 x 3 (76.2 x 76.2)
Coarse	1 (0.30)	1½ (38.1)	1½ x 3 (38.1 x 76.2)
Coarse	5 (1.52)	2 (50.8)	2½ x 3 (63.5 x 76.2)
Coarse	12 (3.66)	2½ (63.5)	3 x 3 (76.2 x 76.2)
Coarse	24 (7.32)	3 (76.2)	3 x 4 (76.2 x 102)

[1]Grout space dimension is the clear dimension between any masonry protrusion and shall be increased by the diameters of the horizontal bars within the cross section of the grout space.
[2]Area of vertical reinforcement not to exceed 6% of the area of the grout space.

(*Specification for Masonry Structures* ACI 530.1/ASCE 6/TMS 602, Masonry Standards Joint Committee, 2002).

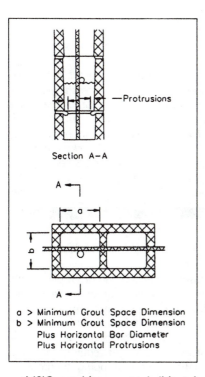

FIGURE 6.11 Grout Space Requirements for Cells or Cores of Hollow Units (Beall and Jaffe, *Concrete and Masonry Databook*, McGraw-Hill, 2003).

MSJC provides a permissible tolerance on the as-constructed dimension of the grout space or cavity width of minus ¼ inch or plus ⅜ inch. This tolerance is not applicable to masonry walls that pass framed construction, because frame tolerances are much higher - usually about plus or minus one inch.

Limitations listed in Table 6.9 can be waived if a grout demonstration panel is used to verify that complete grouting is accomplished despite grouting procedures, construction techniques, and grout space geometry that do not conform to those requirements. The demonstration panel is a mock-up that is erected prior to actual construction using the grouting techniques proposed for the construction.

The maximum grout lift height (height to which grout is placed in one continuous operation) is the smaller of the pour height given in Table 6.9 or five feet. Grout must be consolidated at the time of placement. For grout lifts up to 12 inches high, the grout may be consolidated by mechanical vibration or by puddling (tamping the grout with a bar). Taller grout lifts must be initially mechanically vibrated and then revibrated after initial water loss and settlement

fastfacts

The low-velocity vibrator used for consolidating grout should have a ¾-inch diameter head. Five to ten minutes after the grout is placed, the vibrator should be activated in the grout for one or two seconds in each grouted cell of hollow unit masonry or spaced 12 to 16 inches apart for grout in the collar joint between masonry wythes. Thirty minutes after consolidating the grout, reconsolidate in the same manner. Longer vibration is not only unnecessary, but may damage the masonry by blowing out face shells or separating the wythes.

have occurred. Consolidation prevents voids at the unit-to-grout interface that result from grout shrinkage.

One of two sequences of masonry erection and grout placement may be used: low-lift grouting and high-lift grouting. In low-lift grouting, masonry units are built up to the maximum height permitted by Table 6.9, but no more than five feet. If reinforcement is used, it is placed at this time. The bars must be long enough to project above the grout lift by at least the length required for a lap splice. Grout is then placed into the masonry, stopping the grout 1½ inches below the top of the masonry to form a shear key with the next grout lift, as illustrated in Figure 6.12. Masonry units are then placed for the next pour, and the process is repeated. Low-lift grouting is illustrated in Figure 6.13.

In high-lift grouting, the masonry is erected to its final complete height, but no more than 24 feet. If reinforcement is used, single-piece bars of a length equal to the wall height are placed. Grout is then placed in a 5-foot lift, consolidated and reconsolidated, and then additional grout may be placed. Subsequent grout lifts may be placed on top of the still plastic grout. If so, consolidation of the second lift

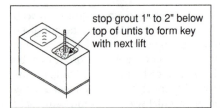

stop grout 1" to 2" below top of untis to form key with next lift

FIGURE 6.12 Grout Shear Key
(Beall and Jaffe, *Concrete and Masonry Databook,* McGraw-Hill, 2003).

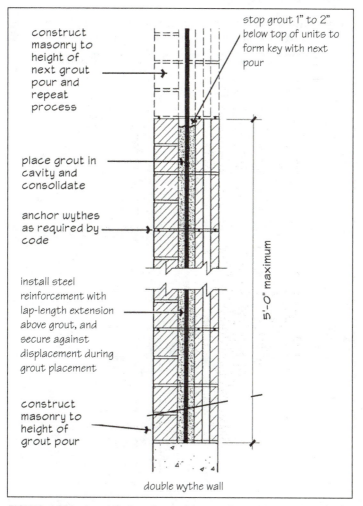

construct masonry to height of next grout pour and repeat process

stop grout 1" to 2" below top of units to form key with next pour

place grout in cavity and consolidate

anchor wythes as required by code

install steel reinforcement with lap-length extension above grout, and secure against displacement during grout placement

construct masonry to height of grout pour

5'-0" maximum

double wythe wall

FIGURE 6.13A Low-Lift Grouting in Masonry Pours Up to 5 Feet High– Double Wythe Wall

(Grouting Masonry, *Masonry Construction Guide* Section 7-11, International Masonry Institute, 1997).

fastfacts

One of two sequences of masonry erection and grout placement may be used: low-lift grouting and high-lift grouting.

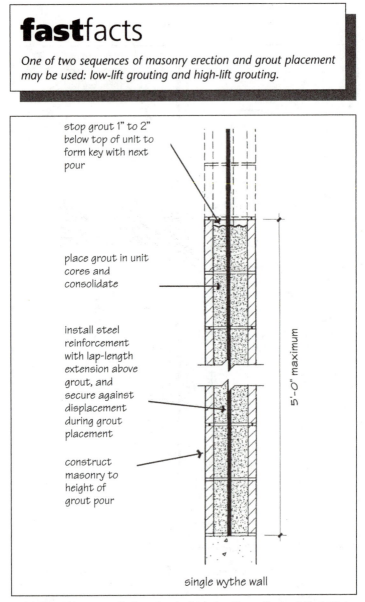

stop grout 1" to 2" below top of unit to form key with next pour

place grout in unit cores and consolidate

install steel reinforcement with lap-length extension above grout, and secure against displacement during grout placement

construct masonry to height of grout pour

5'-0" maximum

single wythe wall

FIGURE 6.13B Low-Lift Grouting in Masonry Pours Up to 5 Feet High– Single Wythe Wall
(Grouting Masonry, *Masonry Construction Guide* Section 7-11, International Masonry Institute, 1997).

includes inserting the vibrator through the lift and into the first lift. If a delay of more than one hour occurs between grout lifts, permitting the first lift to set prior to placement of the subsequent lift, the first lift should be stopped 1½ inches away from a masonry bed joint to prevent formation of a cold joint. High-lift grouting is shown in Figure 6.14.

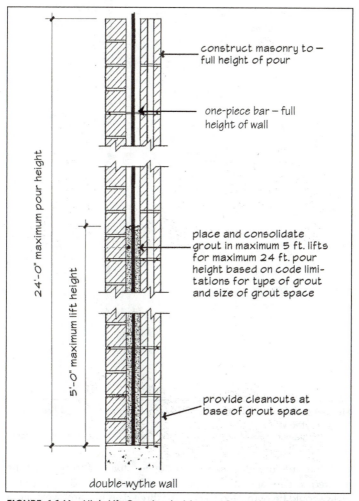

FIGURE 6.14A High-Lift Grouting in Masonry Pours Up to 24 Feet High–Double Wythe Wall
(Grouting Masonry, *Masonry Construction Guide* Section 7-11, International Masonry Institute, 1997)

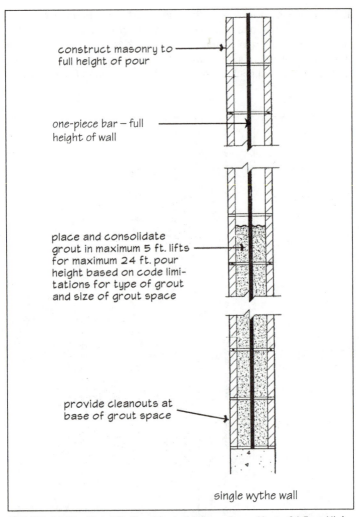

construct masonry to
full height of pour

one-piece bar — full
height of wall

place and consolidate
grout in maximum 5 ft. lifts
for maximum 24 ft. pour
height based on code limitations for type of grout
and size of grout space

provide cleanouts at
base of grout space

single wythe wall

FIGURE 6.14B   High-Lift Grouting in Masonry Pours Up to 24 Feet High–
Single Wythe Wall
(Grouting Masonry, *Masonry Construction Guide* Section 7-11, International Masonry
Institute, 1997)

   Before performing high-lift grouting, the masonry must be
allowed to cure and attain a portion of its strength. Double-wythe
walls should be permitted to cure for three days in warm weather or
five days in cold weather. Single wythe walls should be cured for 12
to 18 hours prior to high-lift grouting.

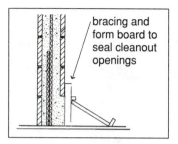

bracing and form board to seal cleanout openings

FIGURE 6.15 Form Closure for Cleanout
(Beall and Jaffe, *Concrete and Masonry Databook*, McGraw-Hill, 2003)

When performing high-lift grouting, openings must be provided at the base of the wall. The openings provide access for removal of mortar droppings from the space that will be grouted. The openings also permit an inspector to observe whether reinforcing bars are correctly positioned and to verify that the space is completely filled after grout placement. These openings, called cleanouts, must be located at each reinforcing bar location or at 32 inches on center. Cleanouts may be formed by omitting masonry units, which are placed after the space is observed and before the grout is placed, or by removing the face shell of selected units in hollow masonry. When the face shell is removed, it may be retained for later installation, or the cleanout may be sealed with form board and bracing as shown in Figure 6.15. Refer to Chapter 7 for proper placement of steel reinforcement.

TABLE 6.10 Approximate Grout Quantities in Collar Joints

Width of Grout Space (in.)	Grout*, cu. yd. of per 100 sq. ft. of Wall	Wall Area, sq. ft. per 1 cu. yd. of grout*
2.0	0.64	157
2.5	0.79	126
3.0	0.96	105
3.5	1.11	90
4.0	1.27	79
4.5	1.43	70
5.0	1.59	63
5.5	1.75	57
6.0	1.91	52
6.5	2.07	48
7.0	2.23	45
8.0	2.54	39

* Includes 3 percent allowance for waste and job conditions.

(Panarese, Kosmatka, & Randall, *Concrete Masonry Handbook*, PCA, 1991).

TABLE 6.11 Approximate Grout Quantities in Hollow Concrete Unit
Masonry[1,2]

Nom. CMU size	Grouted Core Spacing (in. o.c.)	Cu. Yds. Grout per Sq. Ft. of Wall	Cu. Yds. Grout per 100 Units	No. of Units Filled per Cu. Yd. of Grout
6	8	0.93	0.83	120
	16	0.55	0.49	205
	24	0.42	0.37	270
	32	0.35	0.31	320
	40	0.31	0.28	360
	48	0.28	0.25	396
8	8	1.12	1.00	100
	16	0.65	0.58	171
	24	0.50	0.44	225
	32	0.43	0.38	267
	40	0.37	0.33	300
	48	0.34	0.30	330
10	8	1.38	1.23	80
	16	0.82	0.73	137
	24	0.63	0.56	180
	32	0.53	0.47	214
	40	0.47	0.42	240
	48	0.43	0.38	264
12	8	1.73	1.54	65
	16	1.01	0.90	111
	24	0.76	0.68	146
	32	0.64	0.57	174
	40	0.57	0.51	195
	48	0.53	0.47	215

[1]Table assumes horizontal bond beams at 48 inches on center, standard 2-cell CMU of nominal size 8 in. by 16 in., and includes a 3 percent allowance for grout waste and job conditions.
[2]For open-end block, increase quantities by 10 percent. For slump block, reduce quantities by 5 percent.

(Amrhein, *Reinforced Masonry Engineering Handbook,* Masonry Institute of America, 1992).

Quantities of grout that will be required for the construction can be estimated from Tables 6.10 and 6.11. Table 6.10 is for grouting in the collar joint of multi-wythe walls, and Table 6.11 relates to grouting in the cores of hollow concrete unit masonry.

chapter 7

PLACEMENT OF REINFORCEMENT, TIES/ANCHORS, AND FLASHING

Steel reinforcement is added to masonry construction to enhance its strength and ductility. Although strong in compression, masonry is weak in tension. The addition of steel reinforcement increases the masonry's tensile strength. The added strength increases the ability of the masonry to resist applied loads from gravity, wind, and earthquake, but also to resist internal stresses that occur as a result of environmental changes. Reinforcement can also change the failure mode of the masonry from a sudden, brittle failure to a ductile failure in which significant deformations give warning of the impending failure. Of course, the reinforcement must be properly placed in order to achieve these behaviors. Chapter 3 reviewed the material property requirements for steel reinforcement. This chapter presents the placement requirements.

Ties and anchors are part of the connector category, which also includes fasteners. Ties are used to connect together multiple wythes of masonry. Anchors are used to connect masonry to the structural system for lateral support or to connect structural framing components to load-bearing masonry. Fasteners are used to attach non-structural components, such as windows, to masonry. Joint reinforcement can be used as structural reinforcement or as ties.

Masonry walls must not only be able to withstand the code-required applied loads, but must also be able to keep weather out of the enclosed building area. Flashing is an important part of the wall's weather-resistance system. There are a myriad of flashing materials available to the masonry industry, and installation requirements vary somewhat with each type of flashing. A discussion of the advantages and disadvantages of each type of flashing material is outside the scope of this text. This chapter reviews installation requirements for proper functioning of flashing, and assumes that selection of the flashing material has already been made by the designer-of-record.

STEEL REINFORCING BARS

Deformed steel reinforcing bars may be placed in hollow cells or cores of masonry units or between masonry wythes. In either case, the steel bars must be encapsulated in grout. Placement of grout around steel reinforcement is discussed in Chapter 6.

Maximum Reinforcing Bar Size

Just as there are minimum requirements for the size of the space to be grouted, there are limitations on the maximum size of reinforcing bar that can be placed in a masonry member. The maximum reinforcement limitations are imposed by code to ensure adequate flow of the grout around the bars and to permit consolidation of the grout.

The types of maximum reinforcing bar size limitations are based on the bar diameter, the minimum clear dimension of the space into which the bar is placed, the minimum overall thickness of the masonry member that is being reinforced, and the cross-sectional area of the space that will be reinforced and grouted. The requirements are somewhat different for masonry that is designed by allowable stress provisions than for masonry that is designed by strength provisions of the code. Furthermore, there are slight differences between the requirements of the MSJC Code and the IBC. Since only the designer-of-record knows the method by which the masonry was

fastfacts

Just as there are minimum requirements for the size of the space to be grouted, there are limitations on the maximum size of reinforcing bar that can be placed in a masonry member. The maximum reinforcement limitations are imposed by code to ensure adequate flow of the grout around the bars and to permit consolidation of the grout.

structurally designed, that entity must assure that the project documents adhere to these limitations or else must advise the contractor of the appropriate requirements.

For allowable stress design of masonry in accordance with the MSJC Code, the relevant limitations are:

- Reinforcing bar diameter shall not exceed one-half of the minimum clear dimension of the cross-sectional space to be reinforced and grouted, including mortar protrusions, web taper, and other reinforcing bars.

- Reinforcing bar area shall not exceed 6 percent of the cross-sectional area of the space to be grouted.

- Reinforcing bar size shall not exceed No. 11.

The net result of these limitations for reinforcing cores of hollow concrete masonry units is presented in Figure 7.1 and Table 7.1.

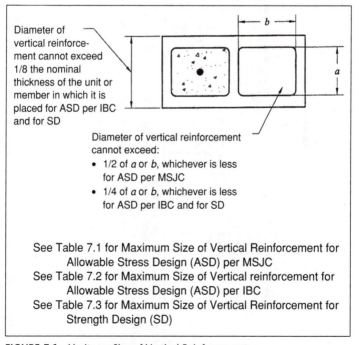

FIGURE 7.1 Limits on Size of Vertical Reinforcement
(Adapted from NCMA, *Annotated Design and Construction Details for Concrete Masonry,* Herndon, VA, 2003).

TABLE 7.1 Maximum Size of Vertical Reinforcement for Allowable Stress Design per MSJC Code

Nominal CMU Thickness[a] (in.)	Cell Size a[b] (in.)	Cell Size b[b] (in.)	Maximum Diameter of Reinforcement (in.)	Maximum Area of Reinforcement (in².)	Maximum Bar Size
			—[c]	—[c]	—[c]
4	0.63	6.19		0.74	No. 7
6	2.13	5.81	1.06	1.26	No. 9
8	3.63	5.81	1.81	1.81	No. 11
10	5.38	5.63	2.68	2.40	No. 11
12	7.13	5.63	2.81		

[a] Typical concrete masonry unit configurations

[b] Refer to Figure 7.1; based on specified minimum dimensions for two-core, square core units and takes into account a cell taper of 1/4 inch and mortar protrusion no larger than 1/2 inch.

[c] Although not limited by Code, grouting of 4-inch concrete masonry units cannot be practically accomplished due to cell size constraints and the resulting inability to adequately consolidate the grout.

(NCMA, *Annotated Design and Construction Details for Concrete Masonry,* Herndon, VA, 2003).

For allowable stress design of masonry in accordance with the IBC, the relevant limitations are:

- Reinforcing bar diameter shall not exceed one-quarter of the minimum clear dimension of the cross-sectional space to be reinforced and grouted, including mortar protrusions, web taper, and other reinforcing bars.
- Reinforcing bar diameter shall not exceed one-eighth of the masonry member thickness.
- Reinforcing bar area shall not exceed 6 percent of the cross-sectional area of the space to be grouted.
- Reinforcing bar size shall not exceed No. 11.

The net result of these limitations for reinforcing cores of hollow concrete masonry units is given in Table 7.2.

The limitations on maximum reinforcing bar size are the same for strength design in accordance with the MSJC Code or the IBC. The relevant limitations are:

- Reinforcing bar diameter shall not exceed one-quarter of the minimum clear dimension of the cross-sectional space to be reinforced and grouted, including mortar protrusions, web taper, and other reinforcing bars.
- Reinforcing bar diameter shall not exceed one-eighth of the masonry member thickness.
- Reinforcing bar area shall not exceed 4 percent of the cross-sectional area of the space to be grouted.
- Reinforcing bar size shall not exceed No. 9.

The net result of these limitations for reinforcing cores of hollow concrete masonry units is given in Table 7.3.

Reinforcing Bar Splice and Hook Requirements

Often it is not possible to place steel reinforcing bar segments that are the full height or length of the masonry element that is to be reinforced. Instead, several pieces of bars are sequentially placed to make up the full reinforced length that is required.

There are three acceptable methods to connect the individual segments of reinforcing bars so that they act together as a single unit. If the bars can be welded (for example, if they meet the requirements of ASTM A 706 as discussed in Chapter 3), the bar

TABLE 7.2 Maximum Size of Vertical Reinforcement for Allowable Stress Design per IBC

Nominal CMU Thickness[a] (in.)	Cell Size a[b] (in.)	Cell Size b[b] (in.)	Maximum Diameter of Reinforcement (in.)	Maximum Area of Reinforcement (in².)	Maximum Bar Size
4	0.63	6.19	—[c]	—[c]	—[c]
6	2.13	5.81	0.53	0.74	No. 4
8	3.63	5.81	0.91	1.26	No. 7
10	5.38	5.63	1.25	1.81	No. 9
12	7.13	5.63	1.41	2.40	No. 11

[a] Typical concrete masonry unit configurations

[b] Refer to Figure 7.1; based on specified minimum dimensions for two-core, square core units and takes into account a cell taper of 1/4 inch and mortar protrusion no larger than 1/2 inch.

[c] Although not limited by Code, grouting of 4-inch concrete masonry units cannot be practically accomplished due to cell size constraints and the resulting inability to adequately consolidate the grout.

(NCMA, *Annotated Design and Construction Details for Concrete Masonry*, Herndon, VA, 2003).

TABLE 7.3 Maximum Size of Vertical Reinforcement for Strength Design

Nominal CMU Thickness[a] (in.)	Cell Size a[b] (in.)	Cell Size b[b] (in.)	Maximum Diameter of Reinforcement (in.)	Maximum Area of Reinforcement (in².)	Maximum Bar Size
4	0.63	6.19	—[c]	—[c]	—[c]
6	2.13	5.81	0.53	0.50	No. 4
8	3.63	5.81	0.91	0.84	No. 7
10	5.38	5.63	1.25	1.21	No. 9
12	7.13	5.63	1.41	1.61	No. 9

[a] Typical concrete masonry unit configurations

[b] Refer to Figure 7.1; based on specified minimum dimensions for two-core, square core units and takes into account a cell taper of 1/4 inch and mortar protrusion no larger than 1/2 inch.

[c] Although not limited by Code, grouting of 4-inch concrete masonry units cannot be practically accomplished due to cell size constraints and the resulting inability to adequately consolidate the grout.

(NCMA, *Annotated Design and Construction Details for Concrete Masonry*, Herndon, VA, 2003).

ends must be butted and welded so as to develop at least 125 percent of the steel's yield strength. Another approach is to provide a mechanical splice (for example, as shown in Figure 3.3). Mechanical splices must also be capable of developing 125 percent of the steel's yield strength. The third method is to overlap the segments of reinforcing bars.

Splicing reinforcing bars by lapping is common practice in both reinforced masonry and reinforced concrete construction. The requirements for length of the overlap are not the same, however, for concrete and masonry. Required reinforcing bar lap lengths are longer in masonry than in concrete, and may be substantially longer. Again, there are some differences in the requirements based on whether the masonry was structurally designed by allowable stress provisions or by strength provisions and based on whether the design is in accordance with the MSJC Code or the IBC.

For masonry that was designed in accordance with the allowable stress provisions of the MSJC Code, the lap splice length requirements are fairly simple: 48 times the nominal bar diameter for Grade 60 reinforcing steel and 40 times the nominal bar diameter for Grade 40 or Grade 50 reinforcing steel. When the bars are coated with epoxy for corrosion protection, these lap lengths must be increased by 50 percent. Table 7.4 lists the required lap lengths for Grade 60 bar sizes No. 3 through No. 11 for allowable stress design per the MSJC Code.

Reinforcing bars that are spliced by lapping need not be in contact. They are permitted to be separated by a distance that is no

TABLE 7.4   Lap Splice Length Requirements for Allowable Stress Design per MSJC Code*

Bar Size, No.	Maximum Transverse Distance Between Lapped Bars, in.	Minimum Lap Splice length, in.
3	3.6	18
4	4.8	24
5	6.0	30
6	7.2	36
7	8.0	42
8	8.0	48
9	8.0	54
10	8.0	60
11	8.0	66

* Based on Grade 60 reinforcing steel. Increase lap lengths by 50 percent when using epoxy-coated reinforcing bars.

(*Specification for Masonry Structures,* ACI 530.1/ASCE 6/TMS 602, Masonry Standards Joint Committee, 1999).

larger than one-fifth the required lap length, but no more than 8 inches. The permitted transverse distances between lapped bars (Grade 60) are also shown in Table 7.4.

For masonry that was designed in accordance with the allowable stress provisions of the IBC or with the strength design provisions of the IBC or the MSJC Code, determination of the required lap splice length is much more complicated. The code provides a formula that considers not only bar diameter and yield stress of the steel but also the specified masonry compressive strength, a bar size factor, and K, where K is the smallest of the masonry cover, the clear spacing between adjacent reinforcement, and five times the nominal bar diameter. For No. 6 bars and larger, this formula results in substantially longer splice length requirements that those that are permitted in allowable stress design in accordance with the MSJC Code. Table 7.5 presents examples of these splice length requirements for the case of specified compressive strength masonry equal to 1,500 psi and Grade 60 reinforcing steel.

A common location in which reinforcing bars are lapped is at the bottom of the wall, where dowels are placed in the concrete foundation and steel reinforcement in the masonry wall is expected to align with those dowels. Often, these dowels are not well aligned with the cells of the hollow masonry units that are placed subsequently. In that case, it is permissible to bend the dowels slightly to get them to fit into the masonry cells. The permitted slope of the bent dowel is 1 inch laterally for each 6 inches vertically, as shown in Figure 7.2.

The ends of reinforcing bars are often bent, or hooked, either to provide lateral support for a perpendicular reinforcing bar or to more

TABLE 7.5 Lap Splice Length Requirements for Strength Design per IBC and MSJC Code, and Allowable Stress Design per IBC*

| Bar Size, No. | Minimum Lap Splice Length, in. | | |
	$K = 5 d_b$	$K = 1.5$ in.	$K = 2$ in.
3	19	24	18
4	25	42	32
5	32	66	49
6	53	132	99
7	62	180	135
8	76	252	189
9	85	320	240

* Based on Grade 60 reinforcing steel and $f'_m = 1500$ psi, where K is the smallest of the masonry cover, the clear spacing between adjacent reinforcement, and 5 d_b.

(NCMA, *Annotated Design and Construction Details for Concrete Masonry,* Herndon, VA, 2003).

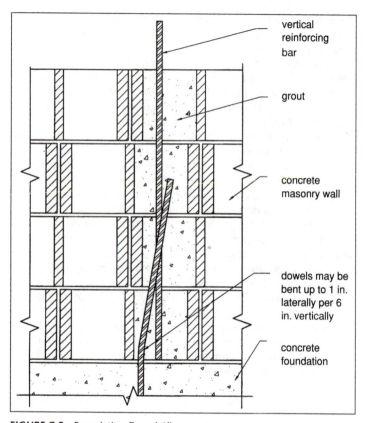

vertical
reinforcing
bar

grout

concrete
masonry wall

dowels may be
bent up to 1 in.
laterally per 6
in. vertically

concrete
foundation

FIGURE 7.2 Foundation Dowel Alignment
(NCMA, *Annotated Design and Construction Details for Concrete Masonry,* Herndon, VA, 2003).

quickly develop the bar's bond with the masonry. Lateral ties in columns and stirrups in beams are hooked reinforcing bars that provide lateral support for the primary reinforcing bars. The stirrup or tie hook has a bend of 90 or 135 degrees, a bend radius of five times the nominal bar diameter, and an extension of six times the nominal bar diameter. Standard hooks are used to more quickly develop primary reinforcing bars. A standard hook has a bend of 90 or 180 degrees and a bend radius, D, that is defined in Table 7.6. The required extension on the 90 degree standard hook is twelve times the nominal bar diameter, and the required extension on the 180 degree standard hook is four times the nominal bar diameter but not less than 2½ inches. Hook requirements are illustrated in Figure 7.3.

TABLE 7.6 Minimum Bend Diameters for Standard Hooks

Bar Size and Type	Minimum Diameter of Bend, D*
No. 3 through No. 7, Grade 40	5 nominal bar diameters
No. 3 through No. 8, Grade 50 or Grade 60	6 nominal bar diameters
No. 9, No. 10, and No. 11, Grade 50 or Grade 60	8 nominal bar diameters

* Refer to Figure 7.3. Does not apply to stirrups or ties.

(NCMA, *Annotated Design and Construction Details for Concrete Masonry,* Herndon, VA, 2003).

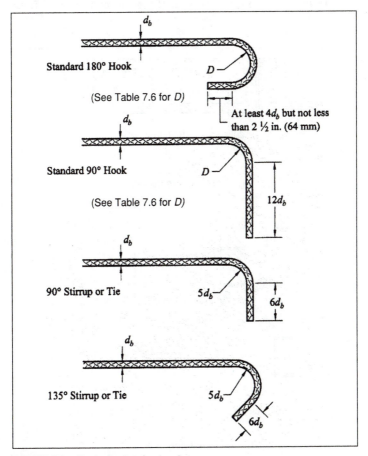

FIGURE 7.3 Hooks for Reinforcing Bars
(NCMA, *Annotated Design and Construction Details for Concrete Masonry,* Herndon, VA, 2003).

Reinforcing Bar Spacing and Cover

When reinforcing bars are placed too close to each other or to the face of the masonry unit, the grout cannot be adequately consolidated completely around the bar. Therefore, the code imposes limitations on these dimensions.

The clear distance between parallel bars in a wall, whether vertical or horizontal, must not be less than the nominal diameter of the bars nor less than 1 inch. For parallel bars in columns and pilasters, the minimum clear distance between bars is increased to not less than 1½ times the nominal diameter of the bars or 1½ inches. These

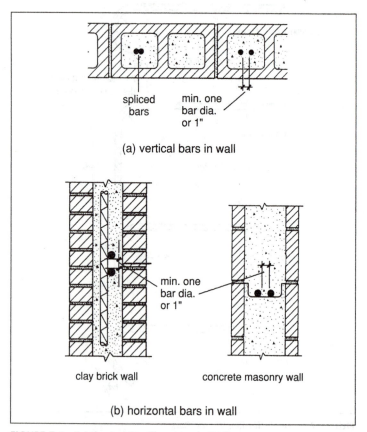

FIGURE 7.4 Reinforcing Bar Clearances in Walls
(Beall & Jaffe, *Concrete and Masonry Databook,* McGraw-Hill, 2003).

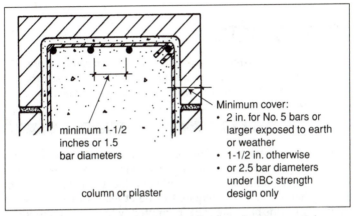

FIGURE 7.5 Bar Spacing and Cover Requirements for Masonry Column or Pilaster
(Beall & Jaffe, *Concrete and Masonry Databook,* McGraw-Hill, 2003).

minimum clearances also apply to contact lap splices and adjacent splices or bars. The minimum clearance between steel reinforcement and the face of the masonry unit depends upon the type of grout that is used around the bar. The minimum clearance when fine grout is used is ¼ inch and the minimum clearance when coarse grout is used is ½ inch. Reinforcing bar clearance requirements are illustrated in Figures 7.4 through 7.6.

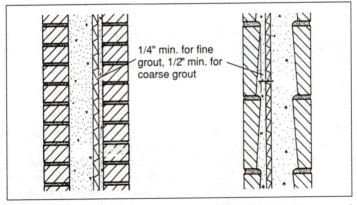

FIGURE 7.6 Minimum Clearance Between Steel Reinforcement and Masonry Units
(Beall & Jaffe, *Concrete and Masonry Databook,* McGraw-Hill, 2003).

fastfacts

Whether or not the reinforcing bar is coated with a corrosion protection product, a minimum amount of distance between the outside face of the masonry construction and the outside face of the steel bar is required by code.

Reinforcing bars are permitted to be bundled in groups of no more than two when the masonry is designed in accordance with the allowable stress provisions. The individual bar cutoff points of bars within a bundle are required to be at least 40 bar diameters apart. Bundled bars are not permitted when the masonry is designed by the strength design method.

Whether or not the reinforcing bar is coated with a corrosion protection product, a minimum amount of distance between the outside face of the masonry construction and the outside face of the steel bar is required by code. For most applications, only the following three parameters apply:

- No. 5 bars and smaller in masonry that is exposed to earth or weather–1½ inches
- Larger than No. 5 bars in masonry that is exposed to earth or weather-2 inches
- Any size bar in masonry that is not exposed to earth or weather-1½ inches

When the masonry is structurally designed by the strength design provisions of IBC, the cover distance is further restricted to no less than 2½ times the nominal diameter of the reinforcing bar.

Tolerances

Accurate placement of reinforcing bars is necessary for the masonry member to achieve its intended strength capacity. Relatively small misalignments in a narrow masonry wall can result in a significant reduction in strength. Therefore, the distance that a reinforcing bar can be placed away from its design position is limited by code. For out-of-plane bending, which pertains to the depth of a bar in a lintel or beam or to the position of a vertical bar in a wall relative to the wall thickness, the permitted tolerances are based on the specified

TABLE 7.7 Placement Tolerances for Reinforcement

Specified Distance d from Face of Wall or Flexural Element to Center of Reinforcement*	Allowable Tolerance for Mild Steel	Allowable Tolerance for Prestressing Steel
$d \leq 8$ in.	± ½ in.	± ¼ in.
8 in. $< d \leq 24$ in.	± 1 in.	± ⅜ in.
$d \geq 24$ in.	± 1¼ in.	± ⅜ in.

*Refer to Figure 7.7

(Specification for *Masonry Structures*, ACI 530.1/ASCE 6/TMS 602, Masonry Standards Joint Committee, 1999).

distance from the face of the wall or beam to the center of the reinforcing bar. The allowable tolerances are listed in Table 7.7 and are graphically shown in Figure 7.7. When the reinforcement consists of prestressing steel rather than mild reinforcing steel, the acceptable tolerances are smaller. The reason for the stricter requirements

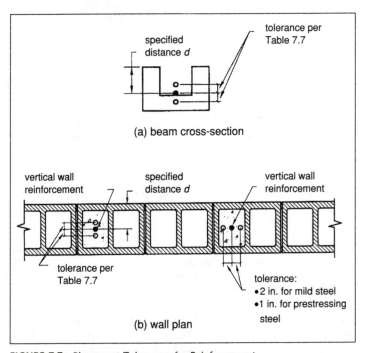

FIGURE 7.7 Placement Tolerances for Reinforcement
(NCMA, *Annotated Design and Construction Details for Concrete Masonry*, Herndon, VA, 2003).

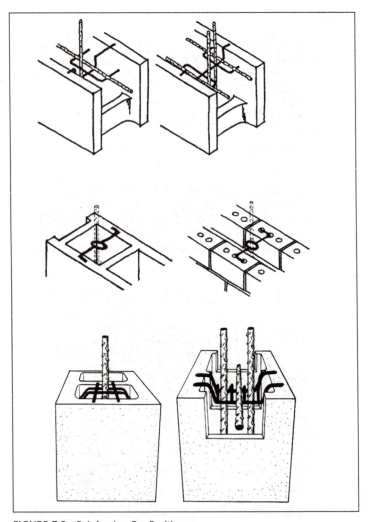

FIGURE 7.8 Reinforcing Bar Positioners
(Beall & Jaffe, *Concrete and Masonry Databook*, McGraw-Hill, 2003 and Panarese, Kosmatka, & Randall, *Concrete Masonry Handbook*, PCA, 1991).

is because tensioned prestressing steel imposes an additional compressive load on the masonry. Misalignment increases the compressive load, which could lead to a sudden, brittle masonry failure.

The tolerance for placement of reinforcement parallel to the length of the wall is more lenient that the tolerances perpendicular to the length of the wall. In this direction, the permitted tolerance is plus or minus 2 inches for mild steel reinforcement and plus or minus 1 inch for prestressing steel reinforcement.

Accessories are available to assist in maintaining the specified placement of reinforcing bars. Reinforcing bar positioners can be used to accurately place both horizontal and vertical reinforcement, either with individual positioners or with accessories that support bars in both directions. Several types of positioners are shown in Figure 7.8. Use of these positioners is not required by IBC or MSJC. These codes instead stipulate a performance requirement of adhering to the tolerance limitations.

PLACEMENT OF JOINT REINFORCEMENT

Joint reinforcement is placed in masonry bed joints between courses of masonry units. It is used to connect together multiple wythes of masonry to help the wall resist lateral loads when the wall spans horizontally, or to help concrete masonry resist tensile stresses that develop as the material dries and shrinks. For each of these uses, the mortar must adequately encapsulate the longitudinal wires in the joint reinforcement in order for the reinforcement to be effective. If the wires are too large relative to the joint width, the wires will not be properly encapsulated. Therefore, the size of the wire that can be placed in a mortar joint is limited to no more than one-half the mortar joint thickness, as illustrated in Figure 7.9.

Various types of joint reinforcement were presented in Chapter 3. The designer-of-record generally selects the appropriate type of joint reinforcement to be used in the construction based on the masonry unit materials that will be used, the type of masonry

fastfacts

It is acceptable to place joint reinforcement directly on the masonry unit surface prior to placing the mortar bed. Studies have shown that the longitudinal wires are sufficiently encapsulated with mortar during normal, subsequent placement of mortar and units.

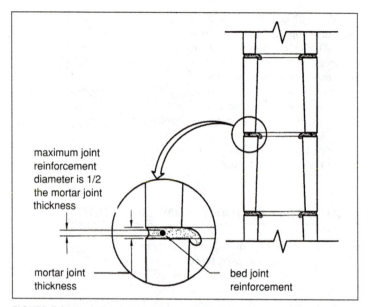

maximum joint
reinforcement
diameter is 1/2
the mortar joint
thickness

mortar joint
thickness

bed joint
reinforcement

FIGURE 7.9 Joint Reinforcement Size Limitation
(NCMA, Annotated *Design and Construction Details for Concrete Masonry*, Herndon, VA, 2003).

assembly that is required (composite or non-composite), and whether or not the wythes in a multi-wythe wall are laid up simultaneously or not. These parameters affect the number of longitudinal wires in the joint reinforcement and the reinforcement type. Joint reinforcement types are ladder, shown in Figure 7.10(a), and truss, seen in Figure 7.10(b).

The number of longitudinal wires should be at least equal to the number of mortar beds on the masonry units. Concrete units are generally laid with two mortar beds, one on each face shell, whether or not the unit is considered solid. Placement of a longitudinal wire in a clay masonry wythe is optional, because clay masonry does not experience overall shrinkage. Therefore, a wall composed of a clay brick outer wythe and a concrete block inner wythe may utilize three longitudinal wires, one in each mortar bed, as shown in Figure 7.11(a), or may employ two longitudinal wires with fixed tab ties (Figure 7.11(c)) or adjustable tab ties (Figure 3.8). When two wythes of concrete masonry make up the wall, the joint reinforcement must contain four longitudinal wires if those units are placed with face shell mortar bedding, as shown in Figure 7.11(b). If those units are

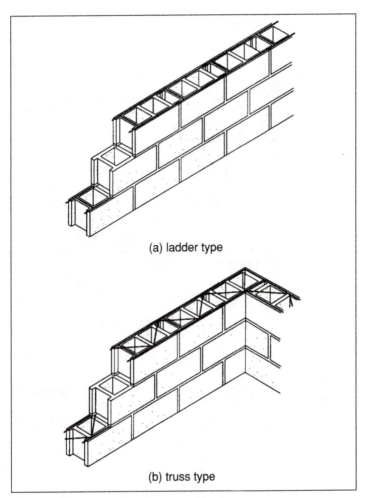

(a) ladder type

(b) truss type

FIGURE 7.10 Basic Joint Reinforcement Types
(Panarese, Kosmatka, & Randall, *Concrete Masonry Handbook*, Portland Cement Association, 1991).

laid with full mortar bedding, as is often done with 4-inch nominal width units, the 4-wire joint reinforcement could still be used, or it could be reduced to 2-wire (for two nominal 4-inch wyths) or 3-wire (for a nominal 4-inch wythe plus a wider wythe) joint reinforcement.

Ladder-type joint reinforcement is always appropriate, but truss-type should not be used when any one of a number of factors apply

---

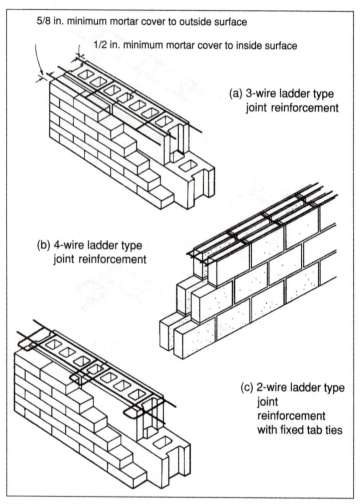

FIGURE 7.11   Longitudinal Wires in Joint Reinforcement
(Panarese, Kosmatka, & Randall, *Concrete Masonry Handbook*, Portland Cement Association, 1991).

to the construction. Factors that contraindicate the use of truss-type joint reinforcement are:

- The masonry units in each wythe of a multi-wythe wall are of different materials.
- The multi-wythe wall will contain insulation in the cavity.

fastfacts

When determining the appropriate type of joint reinforcement to use on a particular project, consideration should also be given to the sequence of construction.

- Vertical reinforcement must be placed in a single-wythe or multi-wythe wall.

When the wythes are of different materials, the intrinsic volume change movement will be in opposite directions. Because truss-type joint reinforcement is relatively stiff axially, it will not permit the differential in-plane movement, and wall bowing could result. When a multi-wythe wall contains insulation in the cavity, the inner wythe is temperature-stable while the outer wythe can experience wide variations in temperature. Thus the inner wythe will not exhibit the same volume changes as the outer wythe. Truss-type joint reinforcement will restrain this differential movement, and wall bowing could result. The last contraindication results from the fact that the diagonal wires in the truss-type joint reinforcement interfere with placement of the reinforcing bars. Furthermore, the MSJC Code prohibits the use of truss-type joint reinforcement in walls that are designed to be non-composite by the allowable stress provisions.

When determining the appropriate type of joint reinforcement to use on a particular project, consideration should also be given to the sequence of construction. When both wythes of a multi-wythe wall are laid up simultaneously, joint reinforcement with longitudinal wires in both wythes can be used, or alternatively a joint reinforcement with longitudinal wires in the concrete masonry wythe only and fixed tab ties into the clay wythe can be used. However, these types should not be used when one wythe is not placed until after the first wythe is completed. In that case, a joint reinforcement with adjustable ties (refer to Figure 3.8) should be used.

For continuity, joint reinforcement should be overlapped at least 6 inches at splices. This relatively short splice length is sufficient because joint reinforcement develops a bond with the mortar by mechanical interlock at the cross wires. At wall corners and at "T" intersections with perpendicular walls, continuity of the joint reinforcement is attained by the used of prefabricated corners and "T's". Manufacturers will fabricate these sections for any configuration of

fastfacts

The MSJC Code prohibits the use of adjustable ties in multi-wythe masonry when the masonry is designed in accordance with the empirical provisions. Therefore, when the masonry has been empirically designed according to the MSJC Code, the wythes must be laid at the same time. The IBC permits the use of adjustable ties in empirically designed masonry, but decreases the square footage of wall surface area that is attributable to each tie.

joint reinforcement that is being used on the project. Examples are shown in Figure 3.9.

A minimum amount of mortar cover is required between the joint reinforcement longitudinal wire and the outside face of the mortar. The minimum cover requirement is based on two needs: protection from the weather to inhibit corrosion and adequate mortar material to resist push-through of the joint reinforcement acting as a tie. Whether solid or hollow masonry units are used, the minimum cover over joint reinforcement placed in walls that are exposed to earth or weather is $\frac{5}{8}$ inch. The minimum cover requirement is reduced to $\frac{1}{2}$ inch when the masonry is not exposed to weather or earth. Figures 7.12 and 7.13 illustrate the cover requirements for solid and hollow units, respectively.

When joint reinforcement is used to tie masonry wythes together, minimum embedment requirements apply. In bed joints of solid units, the minimum embedment is $1\frac{1}{2}$ inches. The tie must be embedded a minimum of $\frac{1}{2}$ inch into the outer face shell mortar

fastfacts

A minimum amount of mortar cover is required between the joint reinforcement longitudinal wire and the outside face of the mortar. The minimum cover requirement is based on two needs: protection from the weather to inhibit corrosion and adequate mortar material to resist push-through of the joint reinforcement acting as a tie.

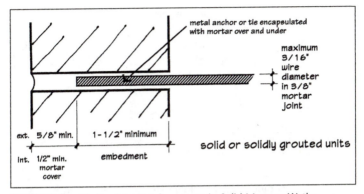

FIGURE 7.12   Placing Joint Reinforcement in Solid Masonry Wythe
(Beall & Jaffe, *Concrete and Masonry Databook,* McGraw-Hill, 2003).

bed of hollow units. These requirements are also shown in Figures 7.12 and 7.13.

Masonry units that are laid in an "other than running bond" pattern are required to have a minimum amount of horizontal reinforcement to provide horizontal continuity. As discussed in Chapter 5, the reinforcement is permitted to consist of joint reinforcement that is placed in the bedding mortar. Table 7.8 lists the maximum spacing of joint reinforcement in "other than running bond" masonry when either standard joint reinforcement (longitudinal wire size W 1.7) or heavy-duty joint reinforcement (longitudinal wire size W 2.8) is used. That table applies to masonry that is empirically

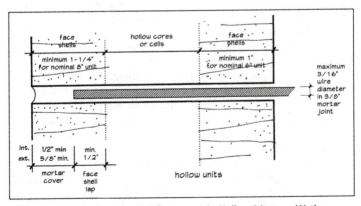

FIGURE 7.13   Placing Joint Reinforcement in Hollow Masonry Wythe
(Beall & Jaffe, *Concrete and Masonry Databook,* McGraw-Hill, 2003).

TABLE 7.8   Maximum Spacing of Joint Reinforcement in "Other Than Running Bond" Masonry

Specified Width of Masonry (in.)	Standard (W 1.7) Joint Reinf. (in.)	Heavy-Duty (W 2.8) Joint Reinf. (in.)
3-5/8	32	48
5-5/8	16	32
7-5/8	16	24
9-5/8	8	16
11-5/8	8	16

(NCMA, *Annotated Design and Construction Details for Concrete Masonry*, Herndon, VA, 2003).

designed or engineered by allowable stress or strength design methods. Masonry veneer that is laid in "other than running bond" is required to have a minimum of one longitudinal wire, of minimum size W 1.7, spaced at no more than 18 inches on center.

FLASHING

Flashings are used in both composite (barrier) wall systems and non-composite (drainage) wall systems to collect water that is inside the wall and provide an avenue for it to drain out of the wall. Although many types of materials are available for use as through-wall flashing, the critical concepts for proper installation of the flashing system are the same, regardless of the material that is specified to be used on the project. The critical concepts include: support of the flashing, termination of the flashing vertical leg, termination of the flashing horizontal leg, flashing splices, flashing dams, flashing corners, and the weep system.

Sheet metal flashings can be installed with a sloped vertical leg in a non-composite or drainage wall, as shown in Figure 7.14(a). The slope promotes drainage of water out of the cavity and out of the wall. Flexible flashings, on the other hand, must be continuously supported. They must be installed to follow the profile of the substrate, as seen in Figure 7.14(b). In a cavity wall, the cavity directly below the flashing must be filled to provide flashing support. If flexible flashings are installed with a slope as shown in Figure 7.14(a), mortar droppings on the flashing will cause the flashing to tear.

The top edge of the vertical leg of the flashing must be terminated in a way that ensures water will not be able to pass behind the flashing. Options available include termination into a reglet, lapping under a moisture barrier system, termination by turning

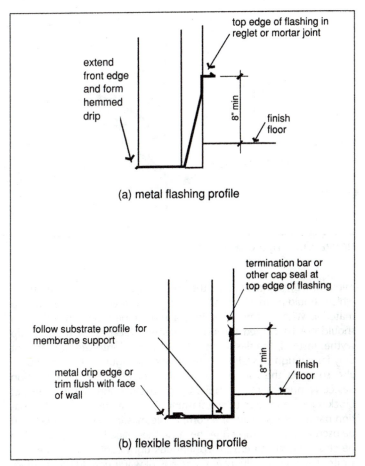

FIGURE 7.14 Flashing Profiles
(Beall & Jaffe, *Concrete and Masonry Databook*, McGraw-Hill, 2003).

into a mortar bed joint in the back-up, and continuous mechanical anchorage and sealant. A reglet is a rectangular slot that is either formed by casting an accessory into the concrete back-up, as shown in Figure 7.15, or by routing out a mortar joint in the masonry back-up. A reglet can be successfully used with sheet metal flashings, but flexible flashings tend to pull out of the reglet under the weight of mortar droppings.

The other options for terminating the vertical leg can be used with any flashing material. The moisture barrier system should lap a

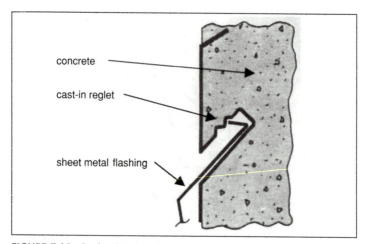

FIGURE 7.15   Reglet Cast into Concrete

minimum of 4 inches over the top of the through-wall flashing, which should be mechanically or adhesively attached to the back-up material. When terminating into a mortar bed joint, the flashing should not be carried through the entire thickness of the back-up wythe. Instead, the flashing should be carried into the mortar joint only far enough to seal the edge, as seen in Figure 7.16. That figure also suggests the use of a cavity filter or other mortar collection device to prevent mortar build-up on the flashing to the extent of blocking water access to the flashing. When a continuous termination bar is used, sealant or a compatible mastic material should also be used to seal the top edge, because the spaced fasteners may not be close enough to ensure continuous tight contact with the substrate. The sealant will prevent passage of water behind the flashing.

Years ago, industry recommendations were to terminate the outside edge, or tip of the horizontal leg of the flashing, ½ inch behind the outside face of masonry. This recommendation probably resulted from aesthetic objections to the flashing projecting out of the wall. However, this practice has proven to be detrimental to the performance of the flashing. Misjudgment of the setback distance and/or retraction of the outside edge under the weight of mortar droppings on the flashing produced many situations where the flashing was far enough back to expose the core holes of the masonry units below, thus rendering the flashing useless. Current industry recommendations are to project the flashing out of the wall.

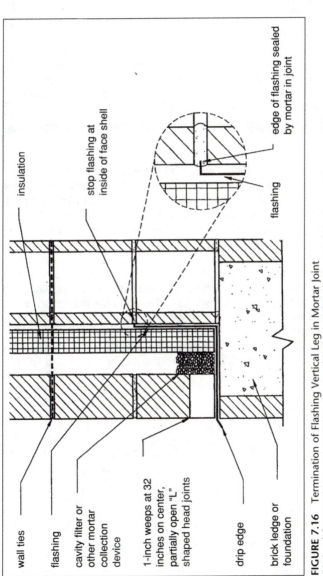

wall ties

flashing

cavity filter or
other mortar
collection
device

1-inch weeps at 32
inches on center,
partially open "L"
shaped head joints

drip edge

brick ledge or
foundation

insulation

stop flashing at
inside of face shell

edge of flashing sealed
by mortar in joint

flashing

FIGURE 7.16 Termination of Flashing Vertical Leg in Mortar Joint
(National Concrete Masonry Association, *Annotated Design and Construction Details for Concrete Masonry*, 2003).

fastfacts

Some manufacturers of self-adhering flashing membranes claim that the adhesive alone is sufficient to hold the vertical leg in place and seal the top edge of the vertical leg of through-wall flashing. In this author's experience, the self-adhesive works well when the flashing is adhered to structural steel members but is inadequate when the flashing must adhere to a masonry substrate.

Sheet metal flashings should be projected out of the wall and bent downward at an angle to form a drip. The outside edge of the flashing should be hemmed, or turned back under, to stiffen the edge and minimize warping. Sealant should be applied below the flashing drip edge only to prevent wind-driven rain from entering the wall below the flashing. Generally, flexible flashings (including composites) cannot be projected out of the wall because these materials degrade with exposure to ultraviolet light. A sheet metal drip edge should be used in conjunction with flexible flashing. The flexible flashing is lapped over and adhered to the drip edge, which projects out of the wall in a similar manner to sheet metal flashing. This system is illustrated in Figure 7.17. The inner edge of the sheet metal drip edge may terminate on the horizontal plane or be bent upward to form a back dam. When the architect-of-record will not

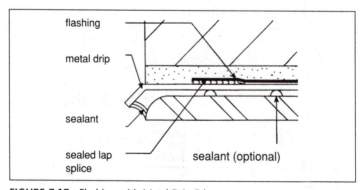

FIGURE 7.17   Flashing with Metal Drip Edge
(Brick Industry Association, *Technical Notes on Brick Construction,* 21B Revised, April 2002).

permit the flashing to extend out of the wall, it should be installed with a projection and cut flush with the outside face of masonry after the construction is complete.

Flexible flashings generally are fabricated in rolls, thus permitting long lengths of walls to be flashed with a single piece. However, the end of the roll occurs frequently enough to make splices in flexible flashing necessary for continuity. Sheet metal flashings are fabricated in finite lengths, often 10 feet, and splices are required in this material as well. Splices should be performed in a way that ensures that a continuous waterproof barrier is produced. General recommendations include lap lengths that are at least 4 inches, use of the manufacturer's recommended adhesive for flexible flashing, and sealing sheet metal flashing with a non-hardening sealant or mastic to prevent water passage. Specific recommendations based on the flashing material type are:

- For metal to metal, solder or seal with high-solids butyl or with self-adhering tape (see Figure 7.18)
- For EPDM, seal with butyl tape or manufacturer's recommended splicing cement
- For self-adhering modified asphalt sheets, apply pressure at seams and apply manufacturer's recommended mastic
- For laminates, seal with manufacturer's recommended mastic
- For metal to EPDM, seal with butyl bonding adhesive
- For metal to modified asphalt sheets, apply pressure at seams and apply manufacturer's recommended mastic

In many locations, lines of flashing are intentionally terminated. Examples include base-of-wall flashing that is terminated at a doorway and lintel flashing that is terminated at each end of the lintel. Flashing terminations must be formed in such a way that water on the flashing cannot travel horizontally into the adjacent masonry. This water path is prevented by formation of end dams. With flexible flashing in a cavity wall, an end dam can be formed by simply folding the flashing, as seen in Figure 7.19. Alternatively, flexible flashings can be cut and lapped to form an end dam as shown in Figure 7.20(a). In sheet metal, flashing end dams can be formed by lapping and soldering per Figure 7.20(b).

Other areas in which discontinuities in the flashing can result if careful attention is not paid to proper installation are inside and outside corners. Continuity can be provided by cutting, lapping, and sealing the ends of flexible flashing or by cutting, lapping, and

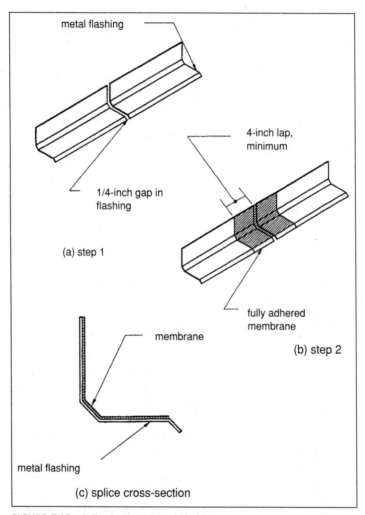

metal flashing

4-inch lap,
minimum

1/4-inch gap in
flashing

(a) step 1

fully adhered
membrane

membrane

(b) step 2

metal flashing

(c) splice cross-section

FIGURE 7.18 Splice in Sheet Metal Flashing
(National Concrete Masonry Association, *Annotated Design and Construction Details for Concrete Masonry,* 2003).

soldering the ends of sheet metal flashing. Details for forming corners in this manner are shown in Figure 7.21. Another approach is to provide separate corner pieces of the same flashing material and to lap and seal them to the continuous flashing. This approach is detailed in Figure 7.22. A third possibility is to provide prefabricated

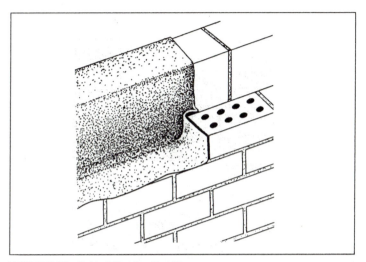

FIGURE 7.19   Folded Flashing End Dam
(Brick Industry Association, *Technical Notes on Brick Construction, 7* Revised, February, 1985).

corner boots, which are then lapped with the continuous flashing as seen in Figure 7.23. Prefabricated corner boots should be installed under rather than over the continuous flashing. Any method that results in continuous, waterproof coverage of both the vertical and horizontal flashing legs over the substrate is acceptable.

Similar to the concept of providing flashing continuity around inside and outside corners, flashing continuity is also required at steps in the flashing. Flashing steps are required at steps in the foundation or adjacent to a sloped roof. One method of providing this continuity is shown in Figure 7.24.

Proper installation of flashing around a window opening requires careful attention to the sequencing of application of flashing materials and moisture barrier. Flashing should be installed not only at the head and sill of the window but also along the jambs. Appropriate sequencing starts with flashing installation at the sill over the top of the moisture barrier system. The jambs are flashed next, and the head is flashed last. The recommended steps in providing window flashing are shown in Figure 7.25.

The flashing system is not complete without provision for weep holes. Weep holes direct water that is collected on the flashing to the exterior of the building. They should always be placed directly on top of the flashing, because that is where the water is collected. Weep holes types are generally divided into small and large openings.

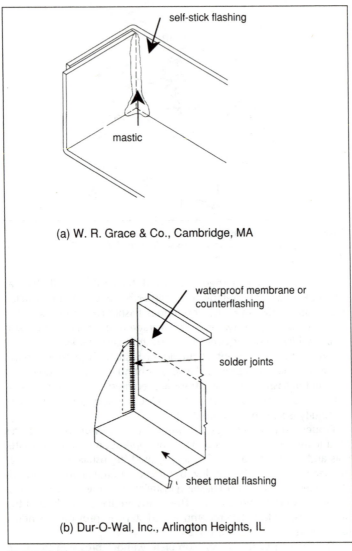

(a) W. R. Grace & Co., Cambridge, MA

(b) Dur-O-Wal, Inc., Arlington Heights, IL

FIGURE 7.20 Flashing End Dams

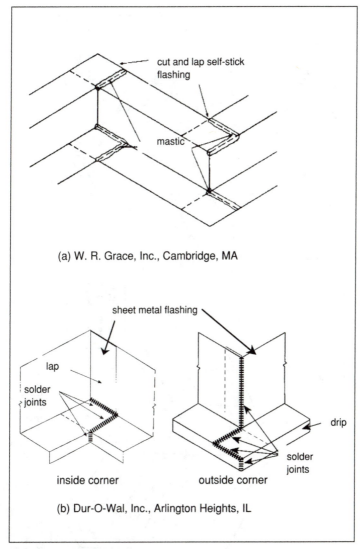

(a) W. R. Grace, Inc., Cambridge, MA

(b) Dur-O-Wal, Inc., Arlington Heights, IL

FIGURE 7.21 Flashing Corners

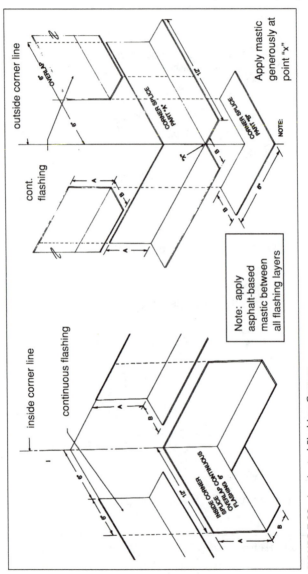

FIGURE 7.22   Laminated Flashing Corners (York Manufacturing, Inc., Sanford, Maine).

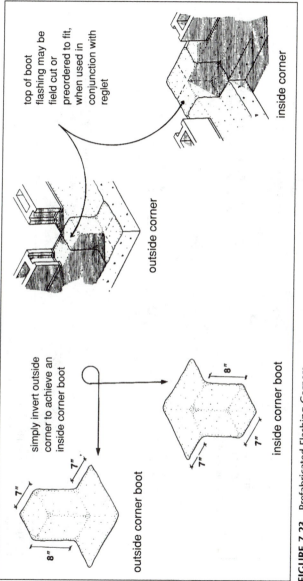

top of boot flashing may be field cut or preordered to fit, when used in conjunction with reglet

outside corner

inside corner

simply invert outside corner to achieve an inside corner boot

7"

7"

8"

outside corner boot

8"

7"

7"

inside corner boot

FIGURE 7.23 Prefabricated Flashing Corners (Illinois Products Corporation, Chicago, IL).

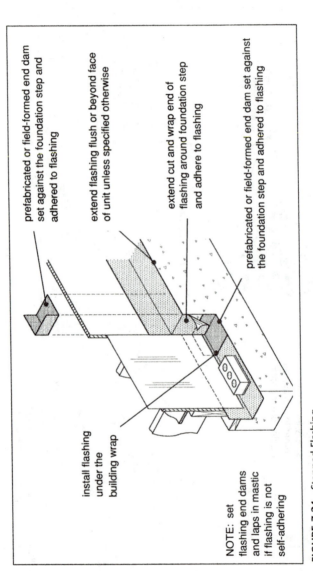

prefabricated or field-formed end dam set against the foundation step and adhered to flashing

extend flashing flush or beyond face of unit unless specified otherwise

extend cut and wrap end of flashing around foundation step and adhere to flashing

prefabricated or field-formed end dam set against the foundation step and adhered to flashing

install flashing under the building wrap

NOTE: set flashing end dams and laps in mastic if flashing is not self-adhering

FIGURE 7.24 Stepped Flashing
(International Masonry Institute, Masonry Construction Guide 8-3.4).

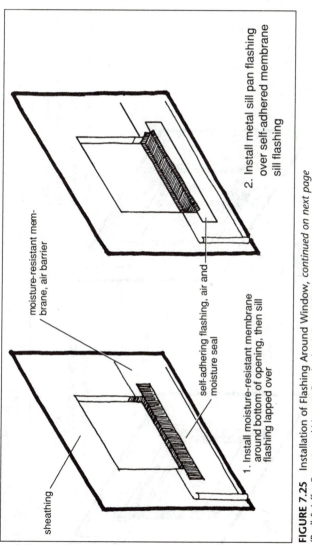

moisture-resistant mem-
brane, air barrier

self-adhering flashing, air and
moisture seal

sheathing

1. Install moisture-resistant membrane
around bottom of opening, then sill
flashing lapped over

2. Install metal sill pan flashing
over self-adhered membrane
sill flashing

FIGURE 7.25 Installation of Flashing Around Window, *continued on next page*
(Beall & Jaffe, *Concrete and Masonry Databook*, McGraw-Hill, 2003).

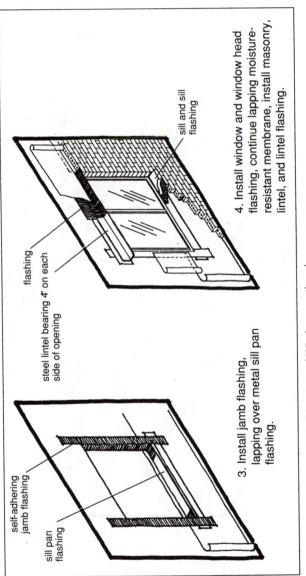

self-adhering
jamb flashing

sill pan
flashing

flashing

steel lintel bearing 4" on each
side of opening

sill and sill
flashing

3. Install jamb flashing,
lapping over metal sill pan
flashing.

4. Install window and window head
flashing, continue lapping moisture-
resistant membrane, install masonry,
lintel, and lintel flashing.

FIGURE 7.25 Installation of Flashing Around Window, *continued*
(Beall & Jaffe, *Concrete and Masonry Databook*, McGraw-Hill, 2003).

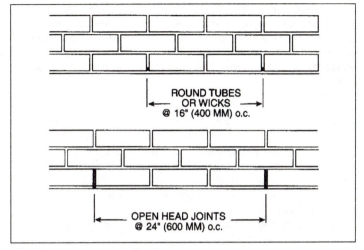

FIGURE 7.26   Weep Hole Spacing
(BIA Technical Notes on Brick Construction 21B, April 2002).

Small weep holes result from the use of plastic tube weeps or rope wicks. Plastic tubes are not recommended because they are easily plugged and because of the small opening that is accessible to the water path. Rope wicks may be left in place if made of an absorbent material or pulled out after the construction is complete, leaving an open hole. Rope wicks that are intended to be pulled out are greased prior to installation. When small weep holes are used, the spacing of the holes should not exceed 16 inches on center.

Large weep holes may be formed by completely omitting mortar from clay brick masonry head joints or by partially omitting mortar from concrete block head joints. Special louvered accessories are available to be placed into open head joints in clay brick masonry. These accessories reduce the risk of insect infestation while maintaining the opening for water egress from the flashed area. The spacing of large weep holes should not exceed 24 inches on center. Figure 7.26 illustrates the recommended spacing of weep holes.

TIES AND ANCHORS

Ties are used to connect together masonry wythes. Code requirements for the permitted types, sizes, and spacing of ties vary depending on the philosophy used to structurally design the masonry (allowable stress design or empirical) and on whether the wall is composite or non-composite. When the masonry is designed by the veneer provisions of the code, anchors are used to connect the veneer to its backing. Anchors may also used at wall intersections (discussed in Chapter 5) and to make connections to the floor or roof diaphragm.

Masonry Ties

In walls that are engineered using the allowable stress design method or that are empirically designed, only wire ties or joint reinforcement is permitted to be used as ties. Adjustable wire ties are only permitted in non-composite walls designed by allowable stress provisions.

Specific requirements for wire ties in composite walls designed by the allowable stress method are:

- Maximum horizontal spacing–36 inches
- Maximum vertical spacing–24 inches
- Maximum wall area per W1.7 wire tie–2⅔ square feet
- Maximum wall area per W2.8 wire tie–4½ square feet
- Joint reinforcement permitted
- Rectangular ties permitted (see Figure 7.27)
- Z-shaped wire ties (see Figure 7.27) permitted only with non-hollow units

In non-composite walls designed by the allowable stress philosophy, the same ties as are permitted in composite walls are allowed with three restrictions: the joint reinforcement must be ladder or tab type; cavity drips are not permitted on the ties; and the maximum cavity width is 4½ inches. Adjustable ties, shown in Figure 7.28, are also permitted in non-composite walls designed by the allowable stress method. The requirements for adjustable ties are:

- Minimum wire size–W2.8
- Maximum horizontal spacing–16 inches
- Maximum vertical spacing–16 inches
- Maximum wall area per tie–1.77 square feet
- Maximum misalignment of bed joints from one wythe to the other–1¼ inches

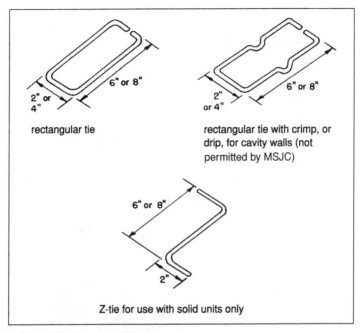

rectangular tie

rectangular tie with crimp, or drip, for cavity walls (not permitted by MSJC)

Z-tie for use with solid units only

FIGURE 7.27 Fixed Unit Ties
(Panarese, Kosmatka, & Randall, *Concrete Masonry Handbook,* Portland Cement Association, 1991).

- Maximum clearance between connecting parts—$\frac{1}{16}$ inch
- Minimum number of pintle legs—2

Additional unit ties must be provided at the perimeter of all openings that are larger than 16 inches in either direction. The maximum permitted spacing of those ties is 3 feet, and the maximum distance from the edge of the opening is 12 inches. Unit ties are also required within 12 inches of unsupported edges of the masonry at a maximum of 24 inches on center vertically or a maximum of 36 inches on center horizontally.

Veneer Anchors

Anchor types that are permitted to be used with masonry veneer that is designed in accordance with the veneer provisions of the MSJC Code are: corrugated sheet metal anchors, sheet metal

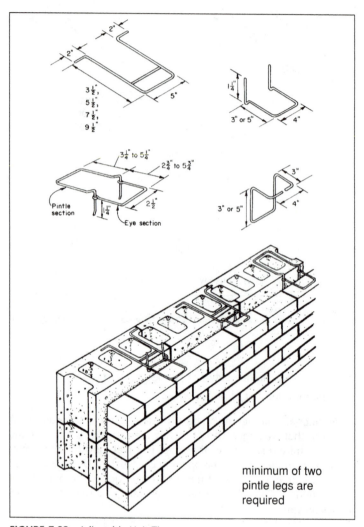

minimum of two
pintle legs are
required

FIGURE 7.28 Adjustable Unit Ties
(Panarese, Kosmatka, & Randall, *Concrete Masonry Handbook,* Portland Cement Association, 1991).

TABLE 7.9 Veneer Anchor Fabrication Requirements

Type	Requirements
corrugated sheet metal	minimum ⅞ in. wide; base metal thickness at least 0.03 in.; corrugations with wavelength of 0.3 to 0.5 in. and amplitude of 0.06 to 0.10 in.
sheet metal	minimum ⅞ in. wide; base metal thickness at least 0.06 in.; either have corrugations defined above or be bent, notched or punched to provide equivalent performance in pull-out or push-through
wire	minimum wire size W1.7; have bent ends to form an extension from the bend of at least 2 in.
joint reinforcement	ladder type or tab type; minimum wire size W1.7; maximum 16 in. spacing of cross wires
adjustable	Meet the requirements for corrugated sheet metal, sheet metal, or wire anchors; meet the requirements for joint reinforcement when applicable; at least two pintle legs of wire size W2.8 with offset not exceeding 1¼ in.

(*Building Code Requirements for Masonry Structures*, ACI 530/ASCE 5/TMS 402, Masonry Standards Joint Committee, 1999 and 2002).

anchors, wire anchors, joint reinforcement, or adjustable anchors. Fabrication requirements for these anchors are given in Table 7.9.

Requirements for spacing of veneer anchors are:

- Maximum horizontal spacing–32 inches
- Maximum vertical spacing–18 inches
- Maximum wall area per adjustable anchor, wire anchor of size W1.7, or 0.03 inch thick (22 gauge) corrugated sheet metal anchor–2.67 square feet
- Maximum wall area per anchor of types other than those listed above–3.5 square feet

The MSJC Code states the thickness requirements for veneer anchors in inches. To convert inch thickness to gauge thickness, refer to Table 7.10. Additional veneer anchors must be provided at the perimeter of all openings that are larger than 16 inches in either direction. The maximum permitted spacing of those

TABLE 7.10   Sheet Metal Sizes

Gauge	Thickness (in.)	Weight (oz/sq.ft.)
10	0.1345	90
12	0.1046	70
14	0.0747	50
16	0.0598	40
18	0.0478	32
20	0.0359	24
22	0.0299	20
24	0.0239	16
26	0.0179	12
30	0.0149	10

(Beall & Jaffe, *Concrete and Masonry Databook,* McGraw-Hill, 2003).

anchors is 3 feet, and the maximum distance from the edge of the opening is 12 inches.

Only certain types of veneer anchors are permitted to be used with each type of permitted backing material. When the veneer is backed by wood stud framing, any anchor type is permitted. When steel stud framing or concrete back the veneer, adjustable anchors must be used. With a masonry back-up for the veneer, wire anchors, adjustable anchors, or joint reinforcement may be used.

Corrugated sheet metal anchors are commonly used with wood stud backing for veneer. The MSJC Code requires a heavier gauge than was traditionally used in residential construction, however. When installing these anchors, care must be taken to install the fastener no more than ½ inch from the bend in the anchor. Anchor effectiveness is greatly reduced as the fastener distance from the bend increases, as illustrated in Figure 7.29.

Requirements for masonry cover over ties and veneer anchors and for embedment of the ties and veneer anchors are the same as those for joint reinforcement. Those requirements are illustrated in Figures 7.12 and 7.13.

Anchor Bolts

Anchor bolt types that are recognized by the MSJC Code are headed bolts, plate bolts, and bent bolts. Anchor bolts for masonry are illustrated in Figure 7.30.

The steel that is used to make anchor bolts conforms to ASTM A 307 "Standard Specification for Carbon Steel Bolts and Studs,

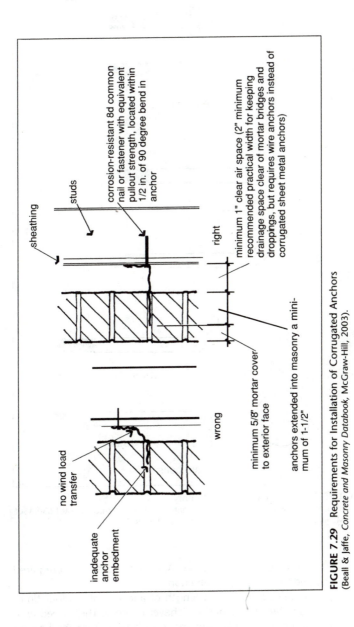

sheathing

studs

corrosion-resistant 8d common nail or fastener with equivalent pullout strength, located within 1/2 in. of 90 degree bend in anchor

right

minimum 1" clear air space (2" minimum recommended practical width for keeping drainage space clear of mortar bridges and droppings, but requires wire anchors instead of corrugated sheet metal anchors)

anchors extended into masonry a minimum of 1-1/2"

minimum 5/8" mortar cover to exterior face

wrong

no wind load transfer

inadequate anchor embedment

FIGURE 7.29   Requirements for Installation of Corrugated Anchors (Beall & Jaffe, *Concrete and Masonry Databook,* McGraw-Hill, 2003).

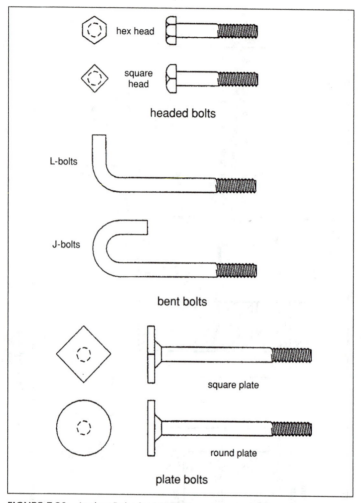

FIGURE 7.30   Anchor Bolts for Masonry
(Commentary on Building Code Requirements for Masonry Structures, ACI 530/ASCE 5/TMS 402, *Masonry Standards Joint Committee,* 1999 & 2002).

60,000 psi Tensile Strength," Grade A. Anchor bolts are not required to be coated for corrosion protection.

The minimum embedment length of anchor bolts is four times the bolt diameter or 2 inches, whichever is larger. The embedment length of plate and headed anchors is measured from the bearing

TABLE 7.11 Anchor Bolt Embedments and Edge Distances

Anchor Bolt Diameter, in.	Minimum Embedment, in.	Edge Distance for Full Shear Capacity, in.
¼	2	3
⅜	2	4½
½	2	6
⅝	2½	7½
¾	3	9
⅞	3½	10½
1	4	12
1¼	5	15

(*Building Code Requirements for Masonry Structures,* ACI 530/ASCE 5/TMS 402, Masonry Standards Joint Committee, 1999 and 2002).

surface of the plate or head perpendicularly to the masonry surface. For bent anchor bolts, the embedment length is measured from the bearing surface of the bent end perpendicularly to the masonry surface except that one bolt diameter is subtracted.

For full capacity in shear, the distance from the center of the bolt to the edge of the masonry in the direction of the shear must be at least twelve bolt diameters. For distances less than this, shear capacity is reduced until the capacity is zero at an edge distance of 1 inch. Table 7.11 presents the values for the minimum embedment length and the edge distance required for full shear capacity of common sizes of anchor bolts in masonry.

chapter

MASONRY CONSTRUCTION DURING WEATHER EXTREMES

Provisions for masonry construction under high and low temperatures permit appropriate strength development and prevent damage to the masonry being placed. Included are special precautions during hot weather and during cold weather.

MASONRY CONSTRUCTION DURING HOT WEATHER

The ideal temperature range for constructing masonry is 60°F (15.5°C) to 80° F (26.7°C). As the ambient temperature increases, the relative humidity at the masonry surface decreases and the evaporation rate increases. During above-normal temperatures (90°F or 32.2°C), these conditions can lead to dry-out of the mortar and grout. Dry-out, or excessive loss of moisture from the mortar or

fastfacts

Dryout is easy to detect. Just scratch the surface of the mortar with a six-penny nail. If mortar material is easily removed from the joint, dry-out has occurred.

grout mix leaving inadequate moisture to hydrate the cement, results in reduced strength and reduced weather resistance. An additional challenge during hot weather is reduced productivity of the mason contractor as a consequence of paying more attention to personal comfort, additional material handling requirements, and extra masonry protection requirements.

A number of adverse effects result from increased mortar temperature, including:

- Workability is reduced-more water must be added to maintain workability.
- Air-entraining agents are less effective-workability is reduced.
- Initial and final set occur more rapidly-length of time that mortar remains usable after initial mixing is reduced.
- More rapid moisture loss from mortar occurs due to unit suction.
- More rapid moisture loss from mortar occurs due to evaporation.

When workability is reduced, placement of masonry units is more difficult and likelihood of decreased mortar-to-unit bond is increased. Moisture loss decreases the amount of water available to participate in the chemical process that makes cement harden-hydration. If the amount of available water is insufficient, mortar may not develop its normal strength. Evaporation also removes moisture more rapidly from the surface of mortar joints, resulting in weaker mortar at the location that is critical for weather-resistance.

Relative humidity, wind velocity, and sunshine affect moisture loss from masonry materials both during and after construction. The effects of relative humidity are not fully understood. However, masonry construction at 90°F (32.2°C) in the dry air of Arizona is more problematic than construction at 90°F (32.2°C) in Florida's humid climate. Scheduling construction activities to avoid the hottest mid-day period should be considered.

fastfacts

Moisture loss decreases the amount of water available to participate in the chemical process that makes cement harden—hydration. If the amount of available water is insufficient, mortar may not develop its normal strength.

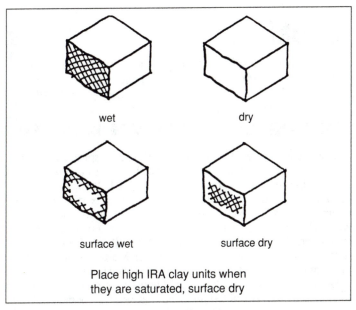

wet   dry

surface wet   surface dry

Place high IRA clay units when
they are saturated, surface dry

FIGURE 8.1   Wetting of High Suction Clay Units
(Beall & Jaffe, *Concrete and Masonry Databook,* McGraw-Hill, 2003).

Other modifications to normal construction procedures include:
reducing the length of mortar spread and elapsed time before placement of units; flushing equipment used to mix, transport, and store
mortar and grout materials immediately before use; and covering
masonry before and after construction. Sand stockpiles should be
sprinkled to increase evaporative cooling and replenish water loss
from evaporation. High-absorption clay units should be pre-wetted to
create a saturated, surface dry condition before placement in the wall.
This moisture condition is illustrated in Figure 8.1. Concrete units
should not be wetted before use or shrinkage will be exacerbated.

fastfacts

*In hot weather, store materials and equipment in the shade to
reduce heat gain from the sun. Temperature reductions of 10°F
(5.6°C) to 30°F (16.7°C) can be realized*

fastfacts

Lower compressive strength mortars should be specified for use in hot weather construction, provided that the structural require-ments are met, because these mortars generally have better water retentivity, board life, and workability.

Lower compressive strength mortars should be specified for use in hot weather construction, provided that the structural requirements are met, because these mortars generally have better water retentivity, board life, and workability. Mortar should be mixed with the maximum amount of water that is consistent with the required workability. Mortar materials should be mixed in a mechanical mixer for the longest period of time consistent with the ASTM C 270 standard (i.e., 5 minutes rather than 3 minutes). Mortar should be retempered as often as needed to maintain workability (unless the mortar is colored), but should be used within 2 hours after initial mixing rather than the 2½ hours permitted under normal weather conditions. Water used to mix mortar and grout should be cool. Ice can be used, but should be completely melted before being added to the mixer.

Admixtures that may be considered for use during hot weather include retarders. Retarders delay the set time of the mortar but do not reduce evaporation rates. Consequently, retarders do not elimi-nate the need for other hot-weather construction provisions that address prevention of too-rapid water loss from the mortar. Air entraining admixtures (those added in the field) are not recom-mended during hot weather, because field control of the addition rate and subsequent mortar air content are difficult to control.

Wind screens of canvas or reinforced polyethylene are effective in reducing the drying effects of wind during hot weather. Fog-spray-ing the masonry for three days after construction may also be ben-eficial. Note that fog-spraying constitutes application of a fine mist of water, not permitting a hose to run down the wall surface. Instead of fog-spraying, the masonry may be completely covered

fastfacts

Wind screens of canvas or reinforced polyethylene are effective in reducing the drying effects of wind during hot weather.

TABLE 8.1 Preparation During Hot Weather Per MSJC 1999 and MSJC 2002

Environmental Conditions	Preparation Prior to Placement
Ambient temperature over 100°F (37.8°C) or Ambient temperature over 90°F (32.2°C) with wind velocity over 8 mph (12.9 km/h)	• Maintain sand piles in damp, loose condition. • Produce mortar with temperature below 120°F (48.9°C).
Ambient temperature over 115°F (46.1°C) or Ambient temperature over 105°F (40.6°C) with wind velocity over 8 mph (12.9 km/h)	• Perform all precautions listed above. • Shade materials and mixing equipment from direct sunlight.

(Specification for Masonry Structures ACI 530.1/ASCE 6/TMS 602, Masonry Standards Joint Committee).

with plastic to retain moisture for curing. As with construction performed during normal temperatures, masonry should be covered at the end of each day to prevent water entry.

Mortar dry-out occurs more frequently in concrete masonry walls than in clay masonry walls because block units should not be wetted (and thereby cooled) prior to placement. Damp curing concrete masonry walls is a preventative measure as well as a curative measure to address dry-out. Damp curing consists of lightly fog-spraying the wall at the end of the work day and covering the wall with plastic. These procedures slow the rate of evaporation from the mortar and increase masonry strength.

Tables 8.1 through 8.3 provide the hot-weather preparation,

TABLE 8.2 Construction During Hot Weather Per MSJC 1999 and MSJC 2002

Environmental Conditions	Precautions During Placement
Ambient temperature over 100°F (37.8°C) or Ambient temperature over 90°F (32.2°C) with wind velocity over 8 mph (12.9 km/h)	• Maintain temperature of mortar and grout below 120°F (48.9°C). • Flush mixer, mortar transport container, and mortar boars with cool water before they contact mortar ingredients or mortar. • Maintain mortar consistency by retempering with cool water. • Use mortar within 2 hours of initial mixing.
Ambient temperature over 115°F (46.1°C) or Ambient temperature over 105°F (40.6°C) with wind velocity over 8 mph (12.9 km/h)	• Perform all precautions listed above. • Use cool mixing water for mortar and grout. Ice is permitted prior to use but not when added to other mortar or grout materials.

(Specification for Masonry Structures ACI 530.1/ASCE 6/TMS 602, Masonry Standards Joint Committee).

TABLE 8.3   Protection During Hot Weather Per MSJC 1999 and MSJC 2002

Environmental Conditions	Protection After Placement
Ambient temperature over 100°F (37.8°C) or Ambient temperature over 90°F (32.2°C) with wind velocity over 8 mph (12.9 km/h)	• Fog spray all newly constructed masonry until damp, at least 3 times a day until the masonry is 3 days old

(Specification for Masonry Structures ACI 530.1/ASCE 6/TMS 602, Masonry Standards Joint Committee).

construction, and protection provisions included in the MSJC Specification. The provisions in the 2002 edition of that standard were unchanged from those in the 1999 edition. These provisions are minimum requirements to prevent dry-out of mortar and grout and to allow for proper curing.

MASONRY CONSTRUCTION DURING COLD WEATHER

Strength development of mortar and grout do not occur when the temperature is below 40°F (4.4°C) or sufficient water is not available. Cement hydration, the chemical process between cement and water that also releases heat, likewise stops below 40°F (4.4°C).

Mortar that has frozen while in the plastic state (prior to hardening) outwardly appears to have hardened. It is able to support loads and bonds to surfaces. However, since water expands when it freezes, mortar may be damaged by the freezing process. Even in the absence of disruptive expansion, frozen mortar has reduced strength. After thawing, once-frozen mortar should be provided with additional water to reactivate the cement hydration process and permit strength gain.

Desirable mortar properties for construction during cold weather include higher strength and lower water retentivity (that is, more water is given up to the masonry unit in suction). Type S mortar is recommended. Mortar materials should be mixed in a mechanical mixer for the shortest period of time consistent with the ASTM C 270 standard (i.e., 3 minutes rather than 5 minutes). Water or sand used in the mortar should be heated to produce mortar temperatures between 40° F (4.4°C) and 120° F (48.9°C). The reason for the lower limit has already been explained. The upper limit is based on the fact that cement will "flash set" at that temperature and be unusable. Use of Type III, high early strength cement, is advantageous in cold weather construction because its rapid strength development balances the slower setting rate that occurs in cold temperatures. Cold weather precautions must still be performed, however, even when Type III cement is used.

When grout freezes, bond to the masonry and reinforcement is reduced. Because of the high water content in grout, expansive forces generated as the water freezes can result in cracking of the masonry units. If Type III, high early strength cement is used in the grout, the protection periods after construction are reduced from 48 hours to 24 hours according to the 2002 MSJC Specification. Other cold weather precautions must still be implemented.

The most commonly used admixtures for mortar during cold weather construction are anti-freezes and accelerators. Some anti-freezes are actually misnomers for accelerators, but true anti-freezes include alcohol. If the quantity of alcohol is sufficient to lower the freezing point of the mortar, compressive and bond strengths are reduced. Therefore, anti-freeze admixtures are not recommended. Accelerator admixtures speed hydration of portland cement. They do not prevent the mortar or grout from freezing. Use of an accelerator may reduce the required protection period but does not eliminate the need for protection or other appropriate cold weather precautions. Calcium chloride is an effective accelerator, but it promotes corrosion of embedded metals in the masonry. It can also contribute to masonry efflorescence and spalling. Consequently, use of calcium chloride is not recommended.

Absorptive masonry units suck water from the mortar into the pores of the unit. This action reduces water in the mortar and reduces the potential for disruptive expansion of water in the mortar while plastic. Conversely, masonry units with low absorptive characteristics do not have the ability to reduce the expansion potential of frozen water in the mortar because they do not remove enough water from the mortar. Consequently, units with high initial rates of absorption are more desirable during cold weather construction. When using units with low absorption rates (initial rate of absorption less than $5g/min/30in^2$), completed construction may need to be protected for longer periods than those recommended by industry standards.

The temperature of the masonry units also has an effect on the freezing potential of the masonry. Excessively cold masonry units, even though dry, may rapidly withdraw heat from the mortar and increase the freezing rate.

The rate at which masonry freezes is dependent upon several factors: mortar temperature and properties; masonry unit temperature and properties; and severity of the temperature and wind.

TABLE 8.4 Preparation During Cold Weather Per MSJC 1999

Ambient Temperature	Preparation Prior to Placement
Below 40° F (4.4°C)	• Heat sand or mixing water to produce mortar temperature between 40°F (4.4°C) and 120°F (48.9°C) at time of mixing.
	• Maintain mortar temperature above freezing until used in masonry.
	• Do not lay masonry units that have temperature below 20°F (−6.7°C).
	• Do not lay masonry units that have visible ice on their surface.
	• Do not lay glass unit masonry when ambient temperature or temperature of the units is below 40°F (4.4°C).

(Specification for Masonry Structures ACI 530.1/ASCE 6/TMS 602, Masonry Standards Joint Committee).

Considerations for the mortar and masonry units have already been discussed. To prevent damage to mortar and grout and mitigate the effects of low temperatures and wind, special precautions are required before, during, and after masonry construction.

In addition to storing materials so that they are kept dry and free from contamination, materials must be heated during cold weather. Water should be heated to produce mortar temperatures between 40° F (4.4°C) and 120° F (48.9°C). When the sand is below 32°F (0°C), it should also be heated. Heating of sand can be accomplished by heaping it over a section of large diameter pipe in which a slow-burning fire is built. Any means of thawing the ice in the sand is acceptable, as long as the sand is not scorched. Mortar and grout temperatures should be maintained above freezing until used in the masonry. If the temperature of the masonry units is below 20°F

TABLE 8.5 Construction During Cold Weather Per MSJC 1999

Ambient Temperature	Precautions During Placement
25° F (-3.9° C) to 20° F (-6.7° C)	• Use heat sources on both sides of masonry under construction.
	• Use wind breaks or enclosures when wind velocity exceeds 15 mph (24 km/h).
20° F (-6.7° C) and below	• Provide an enclosure and auxiliary heat to maintain air temperature above 32° F (0° C) within the enclosure.

(Specification for Masonry Structures ACI 530.1/ASCE 6/TMS 602, Masonry Standards Joint Committee).

TABLE 8.6 Protection During Cold Weather Per MSJC 1999

Mean Daily Temperature	Protection After Placement
Below 40° F (4.4°C)	• Maintain temperature of glass unit masonry above 40°F (4.4°C) for 48 hours.
40° F (4.4°C) to 32° F (0°C)	• Cover completed masonry with weather-resistive membrane for 24 hours.
32° F (0°C) to 25°F (−3.9°C)	• Completely cover completed masonry with weather-resistive membrane for 24 hours.
25° F (−3.9°C) to 20° F (−6.7°C)	• Completely cover completed masonry with insulating blankets or equal for 24 hours.
20° F (−6.7°C) and below	• Maintain completed masonry temperature above 32°F (0°C) for 24 hours by using acceptable auxiliary heating method.

(Specification for Masonry Structures ACI 530.1/ASCE 6/TMS 602, Masonry Standards Joint Committee).

(−6.7°C), they must also be heated. Otherwise, the units will rapidly cool the mortar and grout that come into contact with them.

During cold weather, protection of the masonry immediately after construction generally consists of covering. Depending upon the anticipated temperature range, the masonry should be covered with a weather-resistive membrane or insulating blanket or enclosed and heated. Auxiliary heat is not required unless the anticipated temperature is below 20° F (−6.7°C).

Tables 8.4 through 8.6 summarize the cold weather construction provisions in the 1999 edition of the MSJC Specification. This edition

TABLE 8.7 Preparation During Cold Weather Per MSJC 2002

Ambient Temperature	Preparation Prior to Placement
Below 40°F (4.4°C)	• Do not lay glass masonry. • Do not lay masonry units that have temperature below 20°F (−6.7°C). • Do not lay masonry units that have frozen moisture, visible ice, or snow on their surface. • Remove visible ice and snow from the top surface of foundations and masonry to receive new construction. • Heat foundations and masonry to receive new construction using methods that do not result in damage.

(Specification for Masonry Structures ACI 530.1/ASCE 6/TMS 602, Masonry Standards Joint Committee).

TABLE 8.8   Construction During Cold Weather Per MSJC 2002

Ambient Temperature	Precautions During Placement
40°F (4.4°C) to 32°F (0°C)	• Heat sand or mixing water to produce mortar temperature between 40°F (4.4°C) and 120°F (48.9°C) at time of mixing. • Do not heat sand or water above 140°F (60°C). • Heat grout materials when they are below 32°F (0°C).
32°F (0°C) to 25°F (−3.9°C)	• Heat sand or mixing water to produce mortar temperature between 40°F (4.4°C) and 120°F (48.9°C) at time of mixing. • Maintain mortar temperature above 32°F (0°C) until used in masonry. • Heat grout aggregates and mixing water to produce grout temperature between 70°F (21.1°C) and 120°F (48.9°C) at time of mixing. • Maintain grout temperature above 70°F (21.1°C) until placed in masonry.
25°F (−3.9°C) to 20°F (−6.7°C)	• Perform precautions listed for the temperature range of 32°F (0°C) to 25°F (−3.9°C). • Heat masonry surfaces under construction to 40°F (4.4°C). • Use wind breaks or enclosures when wind velocity > 15 mph (24 km/h).
20°F (−6.7°C) and below	• Perform precautions listed for the temperature range of 32°F (0°C) to 25°F (−3.9°C). • Heat masonry surfaces under construction to 40°F (4.4°C). • Provide an enclosure and auxiliary heat to maintain air temperature above 32°F (0°C) within the enclosure.

(Specification for Masonry Structures ACI 530.1/ASCE 6/TMS 602, Masonry Standards Joint Committee).

is the one referenced in the 2000 International Building Code (IBC). The provisions included in the 2002 edition of the MSJC Specification, which will be referenced in the 2003 IBC, are presented in Tables 8.7 through 8.9.

fastfacts

Overnight, masonry units cool down to the daily low temperature. When construction work starts in the morning, the units may be too cold even if the air temperature has warmed up sufficiently to eliminate the need for the most stringent cold weather precautions. During a day with sunny weather, masonry units usually stay warm enough even if air temperatures indicate that other cold weather precautions are necessary.

TABLE 8.9   Protection During Cold Weather Per MSJC 2002

Temperature*	Protection After Placement
Below 40°F (4.4°C)	• Maintain temperature of glass unit masonry above 40°F (4.4°C) for 48 hours.
40°F (4.4°C) to 25°F (−3.9°C)	• Cover newly constructed masonry with weather-resistive membrane for 24 hours.
25°F (−3.9°C) to 20°F (−6.7°C)	• Cover newly constructed ungrouted masonry with weather-resistive insulating blankets for 24 hours. • Cover newly constructed grouted masonry with weather-resistive insulating blankets for 48 hours. Reduce to 24 hours if the only cement in the grout is Type III portland cement.
20°F (−6.7°C) and below	• Maintain newly constructed ungrouted masonry temperature above 32°F (0°C) for 24 hours by using acceptable auxiliary heating method. • Maintain newly constructed grouted masonry temperature above 32°F (0°C) for 48 hours by using acceptable auxiliary heating method. Reduce to 24 hours if the only cement in the grout is Type III portland cement.

* Anticipated minimum daily temperature for grouted masonry. Anticipated mean daily temperature for ungrouted masonry.

(Specification for Masonry Structures ACI 530.1/ASCE 6/TMS 602, Masonry Standards Joint Committee).

chapter

9

TEMPORARY BRACING OF MASONRY WALLS UNDER CONSTRUCTION

The structural engineer or architect who designs the masonry wall considers the loads that will be imposed on the wall after construction is complete. At that time, the wall is connected to the floor and roof diaphragms or to other elements of the structural frame to provide lateral stability for the wall. Before the wall is complete, it is exposed to wind loading at a time when the masonry is not yet connected to its supporting elements and the masonry does not have its full (28-day) strength. During masonry construction, the wall is most susceptible to damage and even complete failure under high winds. While damage to the work under construction is of concern, the potential for injury or death of workers in the vicinity of masonry under construction is of the highest concern. Bracing requirements are established to improve the ability of free-standing walls to remain standing under certain wind speeds and to provide for the safety of workers under high wind speeds.

fastfacts

During masonry construction, the wall is most susceptible to damage and even complete failure under high winds.

GENERAL BRACING REQUIREMENTS

The MSJC Specification is brief in its treatment of bracing. Part 3 of that Specification states:"Design, provide, and install bracing that will assure stability of masonry during construction." No further instructions are provided to the contractor, who generally is not a structural engineer, as to how to go about designing, providing, and installing bracing that is in compliance. Furthermore, the Specification seems to imply that the bracing should be adequate to prevent the wall from falling no matter how high the wind loads, which is not practical.

In response to the industry need for guidance in masonry bracing, the Council for Masonry Wall Bracing publishes the "Standard Practice for Bracing Masonry Walls Under Construction." The principal goal of that standard is to assure life safety rather than wall stability. While the provisions included in the standard improve the ability of the freestanding wall to withstand wind, it recognizes that walls under construction cannot be prevented from blowing over under very high wind speeds.

Provisions of the Bracing Standard were developed to permit masonry construction to occur during low wind speed conditions without wall bracing. When the wind speed reaches a critical level,

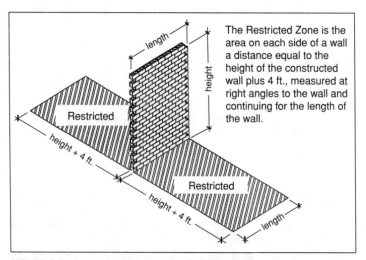

The Restricted Zone is the area on each side of a wall a distance equal to the height of the constructed wall plus 4 ft., measured at right angles to the wall and continuing for the length of the wall.

FIGURE 9.1 Restricted Zone Near Walls Under Construction
(Council for Masonry Wall Bracing, *Standard Practice for Bracing Masonry Walls Under Construction*, Mason Contractors Association of America, Lombard, IL, July 2001).

fastfacts

During the Initial Period, the Restricted Zone on each side of the wall under construction must be evacuated when the observed 5-second wind gust speed exceeds 20 miles per hour as measured by instrument or 15 miles per hour as determined by visual observation.

workers are required to evacuate the dangerous area in the vicinity of the wall under construction. The evacuation area is known as the Restricted Zone, which is illustrated in Figure 9.1. Length of the Restricted Zone is the wall length, and width is equal to the wall height plus 4 feet, measured at right angles to the wall and extending for the specified width from each side of the wall. When the masonry wall reaches the height specified in the Bracing Standard, or when work is finished for the day, the wall is required to be braced.

INITIAL PERIOD OF CONSTRUCTION

The Initial Period is defined as, "the period of time, limited to a maximum of one working day, during which the masonry is being laid above its base or the highest line of bracing and, at the end of which, required bracing is installed."[1] During the Initial Period, the Restricted Zone on each side of the wall under construction must be evacuated when the observed 5-second wind gust speed exceeds 20 miles per hour as measured by instrument or 15 miles per hour as determined by visual observation.

Wind speeds should be determined in close proximity to the wall under construction. The wind direction of concern is that which is normal, or perpendicular, to the wall plane. The permitted wind speed for walls under construction during the Initial Period is increased to 35 miles per hour (30 miles per hour when measured visually) if the masonry is no more than 8 feet above grade. Evacuation wind speeds are listed in Table 9.1.

---

[1] Standard Practice for Bracing Masonry Walls Under Construction, Council for Masonry Wall Bracing, Mason Contractors Association of America, June, 2001.

TABLE 9.1   Restricted Zone Evacuation Wind Speeds

	5-sec. Wind Gust Speed, mph	
Period	Instrument[1]	Visual[2]
Initial	20	15
Intermediate	5 mph less than Design Wind speed	10 mph[3] less than Design Wind speed

[1] May be increased to 35 mph for masonry that is no more than 8 feet above grade.
[2] May be increased to 30 mph for masonry that is no more than 8 feet above grade.
[3] Valid up to 40 mph Design Wind speed; 14% less than the instrumented Evacuation Wind speed above that.

(Council for Masonry Wall Bracing, *Standard Practice for Bracing Masonry Walls Under Construction,* Mason Contractors Association of America, Lombard, IL, July 2001).

The preferred method of measuring wind speed is by instrument. Instruments that measure wind speed and automatically sound an alarm when a set wind speed is reached are commercially available and relatively inexpensive. Instruments should be routinely calibrated to plus or minus 2 miles per hour.

There are a number of methods of visually measuring wind speeds. One such method is the Beaufort Wind Scale, which is listed in Table 9.2. The Bracing Standard requires evacuation of the Restricted Zone at lower wind speeds when measured visually than when measured by instrument because of the subjectivity of visual assessment.

As long as the wind speed stays under the limits given in Table 9.1, work can continue during the Initial Period until the wall

fastfacts

Each project must have a manager who has the authority to evacuate all work forces from the Restricted Zone(s). This person is known as the Authorized Project Agent. The masonry contractor must advise the Authorized Project Agent of the locations of all Restricted Zones. The person who is responsible for determining when the wind speed limitations have been reached must immediately notify the Authorized Project Agent of the need to evacuate the Restricted Zone. The Authorized Project Agent must then ensure that the Zones are evacuated.

TABLE 9.2 Beaufort Wind Scale

No.	Wind Speed	Effects on Land
0	0	Calm: Smoke will rise vertically.
1	1-3	Light Air: Rising smoke drifts but weather vane is inactive.
2	4-7	Light Breeze: Leaves rustle, can feel wind on your face, weather vane is inactive.
3	8-12	Gentle Breeze: Leaves and twigs move around, lightweight flags extend.
4	13-18	Moderate Breeze: Thin branches move, dust and paper are raised.
5	19-24	Fresh Breeze: Trees sway.
6	25-31	Strong Breeze: Large tree branches move, open wires (such as power lines) begin to whistle, umbrellas are difficult to keep under control.
7	32-38	Moderate Gale: Large trees begin to sway, noticeably difficult to walk.
8	39-46	Fresh Gale: Twigs and small branches are broken from trees, walking into the wind is very difficult.

(Council for Masonry Wall Bracing, *Standard Practice for Bracing Masonry Walls Under Construction*, Mason Contractors Association of America, Lombard, IL, July 2001).

height reaches the limits given in Table 9.3. During the Initial Period, mortar is assumed to have no tensile strength, and wall stability is provided only by the self-weight of the masonry materials. Consequently, walls built with heavier units can be built taller before work must be stopped for bracing installation. Solid and grouted units are heavier than hollow units of the same material density, and therefore, walls built with these units can also be built taller than those built with the same density hollow units. Height values listed in the table include a strength reduction factor of 0.9 applied to the stabilizing moment and a load factor of 1.3 applied to the wind pressure.

Permitted unbraced heights given in Table 9.3 apply only when the evacuation wind speed (as measured by instrument) is no more than 20 miles per hour. When the evacuation wind speed exceeds 20 miles per hour, bracing requirements during the Initial Period are the same as those during the Intermediate Period. An exception applies to walls in which the masonry height is no more than 8 feet above grade, for which bracing is not required until wind speed exceeds 35 miles per hour.

TABLE 9.3 Maximum Unbraced Masonry Wall Height During the Initial Period

Maximum Unbraced Height of Masonry Walls Above Grade or Above Highest Line of Lateral Support, Measured on Each Side of the Wall (ft.)

Nominal Wall Thickness (in.)	Lightweight Units (Density 95 to 115 pcf)		Medium Weight Units (Density 115 to 125 pcf)		Normal Weight Units (Density > 125 pcf)	
	Hollow Units	Solid or Solidly Grouted Units	Hollow Units	Solid or Solidly Grouted Units	Hollow Units	Solid or Solidly Grouted Units
4	8'-0"	8'-0"	8'-0"	8'-0"	8'-0"	8'-0"
6	8'-0"	8'-0"	8'-0"	10'-0"	8'-0"	10'-6"
8	10'-8"	15'-4"	12'-8"	18'-0"	14'-0"	20'-0"
10	16'-8"	24'-2"	20'-0"	29'-2"	21'-8"	31'-8"
12	23'-0"	35'-0"	28'-0"	35'-0"	30'-0"	35'-0"

For partially grouted masonry, weight of masonry shall be determined by linear interpolation between hollow ungrouted units and fully grouted units, based on the amount of grouting.

(Council for Masonry Wall Bracing, *Standard Practice for Bracing Masonry Walls Under Construction*, Mason Contractors Association of America, Lombard, IL, July 2001).

INTERMEDIATE PERIOD

The Intermediate Period is defined as, "the period of time following the Initial Period until the wall is connected to the elements that provide its final lateral stability from supporting structural elements."[2] During the Intermediate Period, the Restricted Zone on each side of the wall under construction must be evacuated when the 5-second wind gust speed, as measured by instrument, reaches the design wind speed minus 5 miles per hour or when it is visually determined to be the design speed minus 10 miles per hour (refer to Table 9.1). As long as the wind speed stays under the limits given in Table 9.1, work can continue during the Intermediate Period.

Permitted unbraced wall heights during the Intermediate Period are given in Tables 9.4 through 9.8 for design wind speeds of 40 mph through 20 mph, respectively. Each table lists the allowable unbraced wall heights for 8-inch thick and 12-inch thick masonry units, built with or without reinforcing, and constructed with or without through-wall flashing.

During the Intermediate Period, the masonry assembly is assumed to have half of its 28-day specified compressive strength. When the grout is less than 12 hours old, reinforcement is considered to be ineffective. Between 12 hours and 24 hours of grout age, development lengths and splice lengths of steel reinforcement are required to be 33 percent greater than the minimums given by code for performance under 28-day strengths. At 24 hours or more after placement of grout, the reinforcement is considered to be fully effective at the normal code-required lap and splice lengths. Therefore, the line entries for unreinforced masonry apply to reinforced walls in which the grout is less than 12 hours old.

The Bracing Standard permits masonry bracing to be designed by either the allowable stress approach or by the strength design approach. When using allowable stress design, permitted flexural tensile stress in the masonry is limited to 67 percent of the values given by code for performance under 28-day strengths. Under the strength design approach, permitted flexural tensile stress in the masonry is limited to 40 percent of the modulus of rupture values given by code for performance under 28-day strengths. Note that 40 percent of the modulus of rupture stress is equivalent to the full allowable flexural tensile stress.

Table 9.9 gives requirements for vertical spacing of braces, maximum height of wall above the top brace, and maximum unbraced

---

[2] Ibid.

TABLE 9.4 Maximum Unbraced Wall Height During the Intermediate Period (40 mph Design Wind)

Wall Type	Maximum Unbraced Wall Heights for 40 mph Design Wind Speed*[1]			
	PC-Lime or Mortar Cement Mortar		Masonry Cement Mortar	
	Type M or S	Type N	Type M or S	Type N
8-in., unreinforced, unbonded[2]	3'-4"	3'-4"	3'-4"	3'-4"
8-in., unreinforced, bonded[3]	6'-8"	6'-0"	5'-4"	4'-8"
8-in., reinforced[4], unbonded[2] or bonded[3]				
No. 5 at 10 ft. on center	12'-0"	12'-0"	12'-0"	12'-0"
No. 5 at 4 ft. on center	19'-4"	19'-4"	19'-4"	19'-4"
12-in., unreinforced, unbonded[2]	7'-4"	7'-4"	7'-4"	7'-4"
12-in., unreinforced, bonded[3]	10'-8"	10'-0"	8'-8"	8'-0"
12-in., reinforced[4], unbonded[2] or bonded[3]				
No. 5 at 10 ft. on center	20'-0"	20'-0"	20'-0"	20'-0"
No. 5 at 4 ft. on center	24'-0"	24'-0"	24'-0"	24'-0"

* 35 mph evacuation wind speed by instrument; 30 mph evacuation wind speed visually.
[1] Applies to all hollow concrete masonry of 95 pcf and greater density, all solid concrete masonry, and to hollow and solid clay masonry.
[2] Assumes an unbonded condition between the wall and foundation such as at flashing. Height exception: walls up to 8 ft. tall above the ground do not need to be braced.
[3] Assumes continuity of masonry at the base (i.e., no flashing).
[4] Reinforced walls are considered unreinforced until grout is in place 12 hours. Indicated reinforcement is minimum required and shall be continuous into the foundation. Minimum lap splice for grout less than 24 hours old—40 in.; minimum lap splice for grout 24 hours old or more—30 in. For reinforced walls not requiring bracing, check adequacy of foundation to prevent overturning.

(Council for Masonry Wall Bracing, *Standard Practice for Bracing Masonry Walls Under Construction*, Mason Contractors Association of America, Lombard, IL, July 2001).

TABLE 9.5 Maximum Unbraced Wall Height During the Intermediate Period (35 mph Design Wind)

Wall Type	PC-Lime or Mortar Cement Mortar		Masonry Cement Mortar	
	Type M or S	Type N	Type M or S	Type N
Maximum Unbraced Wall Heights for 35 mph Design Wind Speed*[1]				
8-in., unreinforced, unbonded[2]	4'-0"	4'-0"	4'-0"	4'-0"
8-in., unreinforced, bonded[3]	8'-0"	7'-4"	6'-8"	6'-0"
8-in., reinforced[4], unbonded[2] or bonded[3]				
No. 5 at 10 ft. on center	14'-8"	14'-8"	14'-8"	14'-8"
No. 5 at 4 ft. on center	22'-0"	22'-0"	22'-0"	22'-0"
12-in., unreinforced, unbonded[2]	9'-0"	9'-0"	9'-0"	9'-0"
12-in., unreinforced, bonded[3]	12'-8"	11'-4"	10'-8"	9'-4"
12-in., reinforced[4], unbonded[2] or bonded[3]				
No. 5 at 10 ft. on center	23'-4"	23'-4"	23'-4"	23'-4"
No. 5 at 4 ft. on center	28'-8"	28'-8"	28'-8"	28'-8"

* 30 mph evacuation wind speed by instrument; 25 mph evacuation wind speed visually.

[1] Applies to all hollow concrete masonry of 95 pcf and greater density, all solid concrete masonry, and to hollow and solid clay masonry.

[2] Assumes an unbonded condition between the wall and foundation such as at flashing. Height exception: walls up to 8 ft. tall above the ground do not need to be braced.

[3] Assumes continuity of masonry at the base (i.e., no flashing).

[4] Reinforced walls are considered unreinforced until grout is in place 12 hours. Indicated reinforcement is minimum required and shall be continuous into the foundation. Minimum lap splice for grout less than 24 hours old—40 in.; minimum lap splice for grout 24 hours old or more—30 in. For reinforced walls not requiring bracing, check adequacy of foundation to prevent overturning.

(Council for Masonry Wall Bracing, *Standard Practice for Bracing Masonry Walls Under Construction,* Mason Contractors Association of America, Lombard, IL, July 2001).

TABLE 9.6 Maximum Unbraced Wall Height During the Intermediate Period (30 mph Design Wind)

Maximum Unbraced Wall Heights for 30 mph Design Wind Speed*[1]

Wall Type	PC-Lime or Mortar Cement Mortar		Masonry Cement Mortar	
	Type M or S	Type N	Type M or S	Type N
8-in., unreinforced, unbonded[2]	6'-0"	6'-0"	6'-0"	6'-0"
8-in., unreinforced, bonded[3]	10'-0"	8'-8"	8'-0"	6'-8"
8-in., reinforced[4], unbonded[2] or bonded[3]				
No. 5 at 10 ft. on center	16'-8"	16'-8"	16'-8"	16'-8"
No. 5 at 4 ft. on center	25'-4"	25'-4"	25'-4"	25'-4"
12-in., unreinforced, unbonded[2]	12'-8"	12'-8"	12'-8"	12'-8"
12-in., unreinforced, bonded[3]	15'-4"	14'-0"	12'-8"	11'-4"
12-in., reinforced[4], unbonded[2] or bonded[3]				
No. 5 at 10 ft. on center	28'-8"	28'-8"	28'-8"	28'-8"
No. 5 at 4 ft. on center	33'-4"	33'-4"	33'-4"	33'-4"

* 25 mph evacuation wind speed by instrument; 20 mph evacuation wind speed visually.

[1] Applies to all hollow concrete masonry of 95 pcf and greater density, all solid concrete masonry, and to hollow and solid clay masonry.

[2] Assumes an unbonded condition between the wall and foundation such as at flashing. Height exception: walls up to 8 ft. tall above the ground do not need to be braced.

[3] Assumes continuity of masonry at the base (i.e., no flashing).

[4] Reinforced walls are considered unreinforced until grout is in place 12 hours. Indicated reinforcement is minimum required and shall be continuous into the foundation. Minimum lap splice for grout less than 24 hours old—40 in.; minimum lap splice for grout 24 hours old or more—30 in. For reinforced walls not requiring bracing, check adequacy of foundation to prevent overturning.

(Council for Masonry Wall Bracing, *Standard Practice for Bracing Masonry Walls Under Construction*, Mason Contractors Association of America, Lombard, IL, July 2001).

TABLE 9.7 Maximum Unbraced Wall Height During the Intermediate Period (25 mph Design Wind)

Wall Type	PC-Lime or Mortar Cement Mortar		Masonry Cement Mortar	
Maximum Unbraced Wall Heights for 25 mph Design Wind Speed*[1]	Type M or S	Type N	Type M or S	Type N
8-in., unreinforced, unbonded[2]	8'-8"	8'-8"	8'-8"	8'-8"
8-in., unreinforced, bonded[3]	12'-0"	10'-8"	10'-0"	8'-8"
8-in., reinforced[4], unbonded[2] or bonded[3]				
No. 5 at 10 ft. on center	20'-0"	20'-0"	20'-0"	20'-0"
No. 5 at 4 ft. on center	26'-0"	26'-0"	26'-0"	26'-0"
12-in., unreinforced, unbonded[2]	18'-0"	18'-0"	18'-0"	18'-0"
12-in., unreinforced, bonded[3]	20'-0"	18'-0"	16'-8"	15'-4"
12-in., reinforced[4], unbonded[2] or bonded[3]				
No. 5 at 10 ft. on center	28'-8"	28'-8"	28'-8"	28'-8"
No. 5 at 4 ft. on center	33'-4"	33'-4"	33'-4"	33'-4"

* 20 mph evacuation wind speed by instrument; 15 mph evacuation wind speed visually.

[1] Applies to all hollow concrete masonry of 95 pcf and greater density, all solid concrete masonry, and to hollow and solid clay masonry.

[2] Assumes an unbonded condition between the wall and foundation such as at flashing. Height exception: walls up to 8 ft. tall above the ground do not need to be braced.

[3] Assumes continuity of masonry at the base (i.e., no flashing).

[4] Reinforced walls are considered unreinforced until grout is in place 12 hours. Indicated reinforcement is minimum required and shall be continuous into the foundation. Minimum lap splice for grout less than 24 hours old—40 in.; minimum lap splice for grout 24 hours old or more—30 in. For reinforced walls not requiring bracing, check adequacy of foundation to prevent overturning.

(Council for Masonry Wall Bracing, *Standard Practice for Bracing Masonry Walls Under Construction*, Mason Contractors Association of America, Lombard, IL, July 2001).

TABLE 9.8 Maximum Unbraced Wall Height During the Intermediate Period (20 mph Design Wind)

Maximum Unbraced Wall Heights for 20 mph Design Wind Speed*[1]

Wall Type	PC-Lime or Mortar Cement Mortar		Masonry Cement Mortar	
	Type M or S	Type N	Type M or S	Type N
8-in., unreinforced, unbonded[2]	12'-8"	12'-8"	12'-8"	12'-8"
8-in., unreinforced, bonded[3]	16'-0"	14'-8"	13'-4"	12'-0"
8-in., reinforced[4], unbonded[2] or bonded[3]				
No. 5 at 10 ft. on center	20'-8"	20'-8"	20'-8"	20'-8"
No. 5 at 4 ft. on center	26'-0"	26'-0"	26'-0"	26'-0"
12-in., unreinforced, unbonded[2]	28'-0"	28'-0"	28'-0"	28'-0"
12-in., unreinforced, bonded[3]	27'-4"	25'-4"	23'-8"	22'-0"
12-in., reinforced[4], unbonded[2] or bonded[3]				
No. 5 at 10 ft. on center	28'-8"	28'-8"	28'-8"	28'-8"
No. 5 at 4 ft. on center	33'-4"	33'-4"	33'-4"	33'-4"

* 15 mph evacuation wind speed by instrument; 10 mph evacuation wind speed visually.

[1] Applies to all hollow concrete masonry of 95 pcf and greater density, all solid concrete masonry, and to hollow and solid clay masonry.

[2] Assumes an unbonded condition between the wall and foundation such as at flashing. Height exception: walls up to 8 ft. tall above the ground do not need to be braced.

[3] Assumes continuity of masonry at the base (i.e., no flashing).

[4] Reinforced walls are considered unreinforced until grout is in place 12 hours. Indicated reinforcement is minimum required and shall be continuous into the foundation. Minimum lap splice for grout less than 24 hours old—40 in.; minimum lap splice for grout 24 hours old or more—30 in. For reinforced walls not requiring bracing, check adequacy of foundation to prevent overturning.

(Council for Masonry Wall Bracing, *Standard Practice for Bracing Masonry Walls Under Construction*, Mason Contractors Association of America, Lombard, IL, July 2001).

TABLE 9.9 Bracing Requirements During the Intermediate Period

	Maximum Heights and Spacing[1], ft.-in.			
	PC-Lime or Mortar Cement Mortar		Masonry Cement Mortar	
Bracing Condition	Type M or S	Type N	Type M or S	Type N
8-inch Unreinforced Wall				
Unbraced height, unbonded condition[2]	3'-4"	3'-4"	3'-4"	3'-4"
Height above top brace[3]	6'-8"	6'-0"	5'-4"	4'-8"
Vertical spacing of braces[3]	14'-0"	12'-8"	11'-4"	10'-0"
8-inch Reinforced Wall[4]				
Unbraced height of height above top brace[5]	10'8"	10'8"	10'8"	10'8"
Vertical spacing of braces	21'-4"	21'-4"	21'-4"	21'-4"
12-inch Unreinforced Wall				
Unbraced height, unbonded condition[2]	7'-4"	7'-4"	7'-4"	7'-4"
Height above top brace[3]	10'-8"	10'-0"	8'-8"	8'-0"
Vertical spacing of braces[3]	21'-4"	19'-4"	17'-4"	16'-0"
12-inch Reinforced Wall[4]				
Unbraced height of height above top brace[5]	15'-4"	15'-4"	15'-4"	15'-4"
Vertical spacing of braces	30'-0"	30'-0"	30'-0"	30'-0"

[1] Applies to wall panels up to 25 ft. wide with a brace located at 0.2 times panel width from each end. Applies to all hollow concrete masonry of 95 pcf and greater density, all solid concrete masonry, and to hollow and solid clay masonry.

[2] Assumes an unbonded condition between the wall and foundation such as at flashing. Height exception: walls up to 8 ft. tall above the ground do not need to be braced.

[3] Assumes continuity of masonry other than at the base (i.e., no flashing other than at base).

[4] Reinforced walls shall be considered unreinforced until grout is in place 12 hours. Minimum reinforcement for 8-in. reinforced walls is No. 5 vertical bars at 48 in. on center. Minimum reinforcement for 12-in. reinforced walls is No. 5 vertical bars at 72 in. on center. Minimum lap splice for grout less than 24 hours old—40 in.; minimum lap splice for grout 24 hours old or more—30 in.

[5] Masonry may be bonded or unbonded (i.e., flashing located anywhere in wall). provided that vertical reinforcement is continuous throughout the wall and into the foundation. For reinforced walls not requiring bracing, check adequacy of foundation to prevent overturning.

(Council for Masonry Wall Bracing, Standard Practice for Bracing Masonry Walls Under Construction, Mason Contractors Association of America, Lombard, IL, July 2001).

fastfacts

The Bracing Standard permits masonry bracing to be designed by either the allowable stress approach or by the strength design approach.

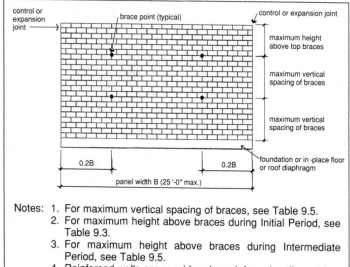

Notes: 1. For maximum vertical spacing of braces, see Table 9.5.
2. For maximum height above braces during Initial Period, see Table 9.3.
3. For maximum height above braces during Intermediate Period, see Table 9.5.
4. Reinforced walls are considered unreinforced until grout has been in place at least 12 hours.

FIGURE 9.2   Required Locations of Temporary Braces
(Adapted from Council for Masonry Wall Bracing, *Standard Practice for Bracing Masonry Walls Under Construction,* Mason Contractors Association of America, Lombard, IL, July 2001).

height during the Intermediate Period. The table assumes that masonry panels are no more than 25 feet wide. Braces are required along the wall horizontally at 0.2 times the panel width from the end. These limitations are illustrated in Figure 9.2. Table 9.9 provisions are applicable to all masonry weighing at least 32 psf (8-inch thick, hollow concrete masonry of 95 pcf density). Table values were based on a design wind speed of 40 mph and are conservative for lower design wind speeds.

BRACING SYSTEMS

Steel, wood, and concrete members of the bracing system are required to be designed in accordance with the relevant code requirements for each material. Bracing systems must be connected to ground anchors or to a dead man (a device that uses a combination of weight and soil pressure to resist brace loads).

The nominal strength of ground anchors is provided by the anchor manufacturer who performs load tests in soils comparable to those at the project site. The design strength of ground anchors is the nominal strength multiplied by a strength reduction factor of 0.6. An additional factor of safety of 2.0 is used when the design is based on unfactored loads.

The nominal strength of dead man anchors is based on the self-weight and soil resistance (if any) only. The design strength of dead man anchors is the nominal strength multiplied by a strength reduction factor of 0.6. An additional factor of safety of 1.5 is used when the design is based on unfactored loads.

Bracing is typically constructed of steel pipes, dimensioned lumber, or steel cables. Typical systems and connections are illustrated in Figures 9.3 through 9.7.

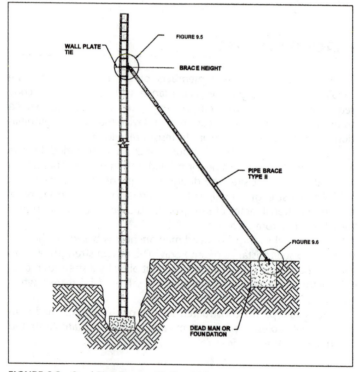

FIGURE 9.3   Steel Pipe Brace with Dead Man
(Dur-O-Wal Technical Bulletin 99-2, 1999).

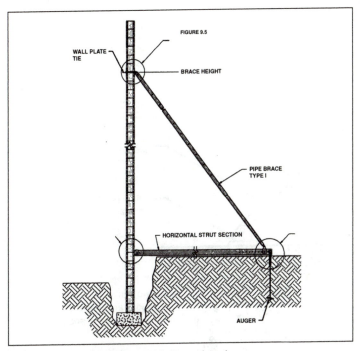

FIGURE 9.4 Steel Pipe Brace with Ground Anchor
(Dur-O-Wal Technical Bulletin 99-2, 1999).

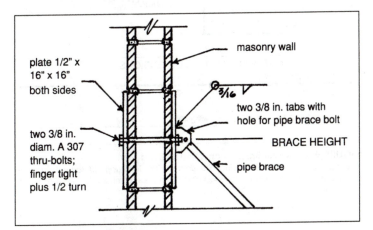

FIGURE 9.5 Typical Brace Connection at Wall
(Council for Masonry Wall Bracing, *Standard Practice for Bracing Masonry Walls Under Construction*, Mason Contractors Association of America, Lombard, IL, July 2001).

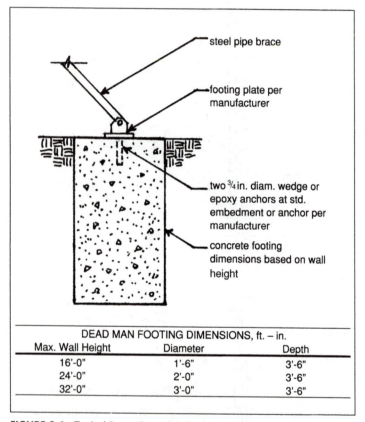

steel pipe brace

footing plate per manufacturer

two ¾ in. diam. wedge or epoxy anchors at std. embedment or anchor per manufacturer

concrete footing dimensions based on wall height

DEAD MAN FOOTING DIMENSIONS, ft. – in.		
Max. Wall Height	Diameter	Depth
16'-0"	1'-6"	3'-6"
24'-0"	2'-0"	3'-6"
32'-0"	3'-0"	3'-6"

FIGURE 9.6 Typical Brace Connection at Dead Man
(Council for Masonry Wall Bracing, *Standard Practice for Bracing Masonry Walls Under Construction*, Mason Contractors Association of America, Lombard, IL, July 2001)

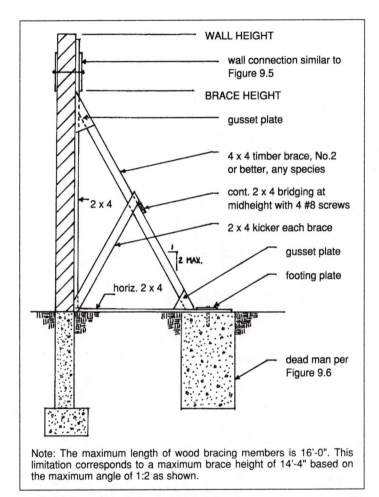

WALL HEIGHT

wall connection similar to Figure 9.5

BRACE HEIGHT

gusset plate

4 x 4 timber brace, No.2 or better, any species

cont. 2 x 4 bridging at midheight with 4 #8 screws

2 x 4 kicker each brace

gusset plate

footing plate

2 x 4

2 MAX.

horiz. 2 x 4

dead man per Figure 9.6

Note: The maximum length of wood bracing members is 16'-0". This limitation corresponds to a maximum brace height of 14'-4" based on the maximum angle of 1:2 as shown.

FIGURE 9.7 Wood Brace with Dead Man
(Council for Masonry Wall Bracing, *Standard Practice for Bracing Masonry Walls Under Construction*, Mason Contractors Association of America, Lombard, IL, July 2001).

GLOSSARY

A

Absorption   the amount of water that a masonry unit absorbs when immersed in cold or boiling water for a specified length of time; expressed as a percentage of the dry unit weight

Accelerator   a liquid or powder ingredient added to a cementitious paste to speed hydration and promote early strength development

Admixture   substances other than water, aggregate, lime, or cement added to mortar or grout to improve one or more chemical or physical properties

Aggregate   granular mineral material such as sand, gravel, crushed stone, or blast furnace slag

Air Entraining Admixture   a powder or liquid formulated to create small, uniformly sized air bubbles when added to a cementitous mix

B

Backup   the part of a multi-wythe masonry wall behind the exterior facing, or the structural portion of the wall to which *veneer* is attached

Bed Joint   horizontal mortar joint between courses of masonry units

Block   a solid or hollow manufactured concrete masonry unit larger than a brick-sized unit

Bond   (1) adhesion between masonry units and mortar or grout, or between steel and mortar or grout (2) connection of masonry wythes or courses by overlapping masonry units

Bond Beam   a horizontal grouted element within masonry in which reinforcement is embedded

Bond Breaker   a material used to prevent adhesion between two surfaces

Bond, Running   placement of masonry units such that head joints in successive courses are horizontally offset at least one-quarter the unit length

Bond, Stack   (1) all masonry not laid in running bond (per *Building Code Requirements for Masonry Structures* ACI 530/ASCE 5/TMS 402) (2) in common use, masonry laid so that head joints in successive courses form a continuous vertical line

C

Cavity Wall   wall composed of two or more wythes of masonry units separated by a continuous air space (exclusive of insulating materials) and in which the wythes are connected by metal ties

Cell (core)   a hollow space within a masonry unit formed by the face shells and webs

Cement   a powdered substance made of burned lime and clay that hardens when mixed with water due to a chemical reaction

Cementitious Material   (1) any inorganic material or mixture of inorganic materials that sets and develops strength by chemical reaction with water (2) for proportioning mortars, the following are considered cementitious materials: portland cement, blended hydraulic cement, masonry cement, mortar cement, lime putty, and hydrated lime

Cleanout   a small opening, typically located in the first course of masonry, for removing mortar droppings prior to grout placement in grouted masonry

Collar Joint   vertical, longitudinal joint, parallel to the face of masonry, located between masonry wythes or between masonry veneer and backing

Composite Wall   multi-wythe wall of similar or dissimilar masonry units, constructed so that the wythes act together as one member to resist applied loads; the wythes must be tied with masonry headers or must have a solid collar joint and metal ties

Control Joint   a continuously formed, sawed, tooled, or assembled joint that creates a weakened plane within the masonry so as to regulate the location and degree of cracking resulting from shrinkage of cementitious products

Corrosion destruction of metal by chemical, electrochemical or electrolytic reaction with its environment or other materials

Course a horizontal layer of masonry units one unit in thickness

Curing maintaining proper conditions of moisture and temperature to increase strength and reduce shrinkage in products containing portland cement

D

Dowel a straight metal bar used to connect masonry to masonry or to concrete

Drip a groove or slot cut beneath and slightly behind the forward edge of a projecting unit or element, such as a sill, lintel, or doping, to cause rainwater to drip off and prevent it from penetrating the wall

E

Efflorescence crystalline deposit, usually white, deposited on the surface of masonry after evaporation of moisture from within the masonry

Exfoliation delamination of stone unit surfaces

Expansion Joint a continuous joint, free of non-compressible materials, that accommodates expansion and differential movement of masonry on each side of the joint

F

Face Shell side wall of a hollow masonry unit

Face Shell Bedding application of mortar to the horizontal and vertical surfaces of face shells only when laying hollow masonry units

Flashing a relatively thin, impermeable material placed in masonry to collect water that penetrates the outside wythe and to direct the water to the building exterior

Full Mortar Bedding application of mortar to the entire horizontal surface when laying masonry units

G

Gradation the particle size distribution of aggregate as determined by separation with standard screens, expressed in terms of the individual percentages passing the screens

Grout a mixture of cementitious materials, aggregates, and water, with or without admixtures, initially produced to a liquid consistency

Grout Lift height to which grout is placed in a cell, collar joint, or cavity without intermission

Grout Pour total height of masonry erected prior to placement of grout in one or more lifts

Grouting, High Lift the technique of grouting masonry in lifts after the wall is constructed to its full height

Grouting, Low Lift the technique of grouting masonry as the wall is constructed, usually to heights of 4 to 6 feet

H

Head top of a window or door opening

Head Joint vertical mortar joint, perpendicular to the face of masonry, between masonry units within the same wythe

Header a masonry unit placed with its length perpendicular to the face of the wall so as to connect two or more adjacent wythes of masonry

Height vertical dimension of masonry or a masonry unit, when measured parallel to the intended face

High Lift Grouting method of grouting masonry only after the entire height of masonry to be grouted has been erected

Hollow Masonry Unit a masonry unit whose cross-sectional area in any plane parallel to the bearing surface is less than 75 percent of its gross cross-sectional area measured in the same plane

I

Initial Rate of Absorption (IRA) a measure of the suction of water into a dry clay masonry unit bed face during one minute of exposure, expressed as grams of water divided by the net area

J

Joint Reinforcement   prefabricated metal wire assembly placed in mortar bed joints to help resist concrete masonry shrinkage or to connect multiple wythes of masonry

L

Lap   (1) the distance two bars overlap when forming a splice (2) the distance one masonry unit extends over another

Lap Splice   the connection between reinforcing steel generated by overlapping the ends of the reinforcement

Length   horizontal dimension of masonry or a masonry unit when measured parallel to the intended face

Lintel   beam placed over an opening in a wall to carry the super-imposed load

Low Lift Grouting   technique of grouting masonry in increments during the course of masonry construction

M

Masonry   assembly of masonry units, set in mortar or grout, with or without steel reinforcement

Masonry Unit   natural or manufactured clay, concrete, stone, or glass of a unit size suitable for construction of masonry

Mortar   mixture of cementitious materials, fine aggregate, and water, with or without admixtures, initially produced to a stiff, sticky consistency

Mortar Bed   a horizontal layer of mortar used to seat a masonry unit

Movement Joint   general term that includes control joints and expansion joints

N

Nominal Dimension   masonry or a masonry unit size stated as the specified dimension plus the thickness of a mortar joint (usually stated in whole numbers)

Non-Composite Wall   multi-wythe wall of similar or dissimilar masonry units, constructed so that the wythes act independently to resist applied loads

P

Parapet Wall   portion of the wall that extends above the roof
Partitition Wall   interior, non-load-bearing wall
Pilaster   portion of a wall that projects on one or both sides, is built integrally with the wall, and acts as a vertical beam, column, architectural feature, or any combination thereof
Plasticizer   a substance incorporated into a cementitious material to increase its workability, flexibility, or extensibililty
Prism   assemblage of masonry units and mortar, with or without grout, used as a test specimen

Q

Quality Assurance   the administrative and procedural requirements established by contract documents and by code to assure that constructed masonry is in compliance
Quality Control   the planned system of activities used to provide a level of quality that meets the needs of the users; the use of such a system

R

Repointing   the combined process of cutting out defective mortar joints and filling them in with fresh mortar
Retarding Agent   a chemical additive in mortar that slows setting or hardening
Retempering   adding more water to mortar after initial mixing, but before partial set has occurred, to restore workability
Rowlock   a brick placed with its long face down and its end surface visible at the wall surface

S

Sand particles of fine aggregate that will pass a No. 4 sieve and be retained on a No. 200 sieve

Shear Wall wall that resists lateral load acting within its own plane

Shelf Angle continuous L-shaped structural steel member placed horizontally and attached to the building structure so as to support the weight of masonry

Shrinkage the volume change due to moisture loss, decrease in temperature or carbonation of a cementitious material

Sill bottom of a window opening

Slump the decrease in height of a wet cementitious material immediately after removal of the mold; a measure of workability

Slushed Joint a mortar joint filled after units are laid by "throwing" mortar in with the edge of a trowel

Solid Masonry Unit a unit whose cross-sectional area in every plane parallel to the bearing surface is 75 percent or more of its gross cross-sectional area measured in the same plane

Spall to flake or split away through frost action or pressure

Specified Dimension dimension to which the masonry unit is required to conform by manufacture

Stretcher masonry unit laid with its greatest dimension horizontal and its face parallel to the wall face

T

Temper to moisten and mix mortar to a proper consistency

Thickness horizontal dimension of masonry or masonry unit, when measured perpendicular to the intended face

Tolerance the specified allowance in variation from a specified size, location, or placement

Tooling compressing and shaping the surface of a mortar joint so as to densify its surface and promote good bond with the adjacent masonry units

V

Veneer   a single facing wythe of masonry units which is attached to the backup by anchors or by adhesion so as to transfer loads

W

Wall, Multi-Wythe   a wall composed of two or more masonry wythes

Wall, Single Wythe   a wall of one masonry unit thickness

Water-Repellent   property of a surface that resists penetration by liquid water.

Waterproof   property of a surface that prevents penetration by liquid water applied under pressure

Weep/Weephole   a small hole in the mortar joint directly above flashing to permit escape of moisture

Workability   ability of mortar to be easily placed and spread while still supporting the weight of masonry units placed on it

Workmanship   (1) the art or skill of a workman (2) craftsmanship; the quality imparted to a masonry element

Wythe   a vertical layer of masonry units one unit in thickness

ACRONYMS

ACI   American Concrete Institute, P. O. Box 9094, Farmington Hills, Michigan 48333, 248-848-3700, www.aci-int.org

ASCE   American Society of Civil Engineers, 1801 Alexander Bell Drive, Reston, Virginia 20191, 800-548-2723, 703-295-6300, www.asce.org

ASTM   ASTM International (formerly American Society for Testing and Materials), 100 Barr Harbor Drive, West Conshohocken, Pennsylvania19428-2959, 610-832-9585, www.astm.org

BIA   Brick Industry Association (formerly Brick Institute of America), 11490 Commerce Park Drive, Reston, Virginia 20191, 703-620-0010, www.bia.org

IBC   International Building Code, International Code Council, Inc., 5203 Leesburg Pike, Suite 708, Falls Church, Virginia 22041-3401, 703-931-4533

MCAA   Mason Contractors Association of America, 33 S. Roselle Road, Schaumburg, Illinois 60193, 800-536-2225, www.masonryshowcase.com

MSJC   Masonry Standards Joint Committee, sponsored by American Concrete Institute, American Society of Civil Engineers, and The Masonry Society, www.masonrystandards.org

MSJC Code   Building Code Requirements for Masonry Structures, ACI 530/ASCE 5/TMS 402, Masonry Standards Joint Committee

MSJC Specification   Specification for Masonry Structures, ACI 530.1/ASCE 6/TMS 602, Masonry Standards Joint Committee

NCMA   National Concrete Masonry Association, 13750 Sunrise Valley Drive, Herndon, Virginia 20171-3499, 703-713-1900, www.ncma.org

PCA   Portland Cement Association, 5420 Old Orchard Road, Skokie, Illinois 60077, 847-966-6200, www.portcement.org

TMS   The Masonry Society, 3970 Broadway, Suite 201-D, Boulder, Colorado 80304-1135, 303-939-9700, www.masonrysociety.org

ASTM STANDARDS REFERENCED IN THIS BOOK

A 82   Specification for Steel Wire, Plain, for Concrete Reinforcement

A 153   Specification for Zinc Coating (Hot-Dip) on Iron and Steel Hardware

A 185   Specification for Steel Welded Wire Fabric, Plain, for Concrete Reinforcement

A 496   Specification for Steel Wire, Deformed, for Concrete Reinforcement

A 497   Specification for Steel Welded Wire Fabric, Deformed, for Concrete Reinforcement

A 615   Specification for Deformed and Plain Billet-Steel Bars for Concrete Reinforcement

A 641   Specification for Zinc-Coated (Galvanized) Carbon Steel Wire

A 706   Specification for Low-Alloy Steel Deformed and Plain Bars for Concrete Reinforcement

A 767   Specification for Zinc-Coated (Galvanized) Steel Bars for Concrete Reinforcement

A 775   Specification for Epoxy-Coated Steel Reinforcing Bars

A 884   Specification for Epoxy-Coated Steel Wire and Welded Wire Fabric for Reinforcement

A 951   Specification for Masonry Joint Reinforcement

A 996   Specification for Rail-Steel and Axle-Steel Deformed Bars for Concrete Reinforcement

C 5   Specification for Quicklime for Structural Purposes

C 55   Specification for Concrete Brick

C 62   Specification for Building Brick (Solid Masonry Units Made From Clay or Shale)

C 67 Test Methods for Sampling and Testing Brick and Structural Clay Tile

C 73 Specification for Calcium Silicate Brick (Sand-Lime Brick)

C 90 Specification for Load-Bearing Concrete Masonry Units

C 91 Specification for Masonry Cement

C 97 Test Methods for Absorption and Specific Bulk Gravity of Dimension Stone

C 99 Test Methods for Modulus of Rupture of Dimension Stone

C 120 Test Methods for Flexure Testing of Slate (Modulus of Rupture, Modulus of Elasticity)

C 121 Test Methods for Water Absorption of Slate

C 126 Specification for Ceramic Glazed Structural Clay Facing Tile, Facing Brick, and Solid Masonry Units

C 129 Specification for No-Load-Bearing Concrete Masonry Units

C 140 Test Methods for Sampling and Testing Concrete Masonry Units and Related Units

C 144 Specification for Aggregate for Masonry Mortar

C 150 Specification for Portland Cement

C 170 Test Methods for Compressive Strength of Dimension Stone

C 207 Specification for Hydrated Lime for Masonry Purposes

C 216 Specification for Facing Brick (Solid Masonry Units Made From Clay or Shale)

C 217 Test Methods for Weather Resistance of Slate

C 241 Test Methods for Abrasion Resistance of Stone Subjected to Foot Traffic

C 270 Specification for Mortar for Unit Masonry

C 404 Specification for Aggregates for Masonry Grout

C 410 Specification for Industrial Floor Brick

C 476 Specification for Grout for Masonry

C 503 Specification for Marble Dimension Stone (Exterior)

C 568 Specification for Limestone Dimension Stone

C 595	Specification for Blended Hydraulic Cement
C 615	Specification for Granite Dimension Stone
C 616	Specification for Quartz-Based Dimension Stone
C 629	Specification for Slate Dimension Stone
C 652	Specification for Hollow Brick (Hollow Masonry Units Made From Clay or Shale)
C 744	Specification for Prefaced Concrete and Calcium Silicate Masonry Units
C 780	Test Method for Preconstruction and Construction Evaluation of Mortars for Plain and Reinforced Unit Masonry
C 880	Test Methods for Flexural Strength of Dimension Stone
C 902	Specification for Pedestrian and Light Traffic Paving Brick
C 936	Specification for Solid Concrete Interlocking Paving Units
C 952	Test Method for Bond Strength of Mortar to Masonry Units
C 979	Specification for Pigments for Integrally Colored Concrete
C 1019	Test Method for Sampling and Testing Grout
C 1072	Test Method for Measurement of Masonry Flexural Bond Strength
C 1088	Specification for Thin Veneer Brick Units Made From Clay or Shale
C 1157	Specification for Hydraulic Cement
C 1272	Specification for Heavy Vehicular Paving Brick
C 1314	Test Method for Compressive Strength of Masonry Prisms
C 1319	Specification for Concrete Grid Paving Units
C 1324	Test Method for Examination and Analysis of Hardened Masonry Mortar
C 1329	Specification for Mortar Cement
C 1353	Test Methods for Using the Tabor Abraser for Abrasion Resistance of Dimension Stone Subjected to Foot Traffic
C 1357	Test Method for Evaluating Masonry Bond Strength

C 1384   Specification for Admixtures for Masonry Mortars
C 1405   Specification for Glazed Brick (Single Fired, Solid Brick Units)

E 514   Test Method for Water Penetration and Leakage Through Masonry
E 518   Test Method for Flexural Bond Strength of Masonry

INDEX